TOP CANADIAN
CONTEMPORARY ARCHITECTS
加拿大当代顶级建筑师

张钰华 编

江苏人民出版社

图书在版编目（CIP）数据

加拿大当代顶极建筑师 / 张钰华编. —— 南京：江苏人民出版社，2012.2
ISBN 978-7-214-07914-5

Ⅰ.①加… Ⅱ.①张… Ⅲ.①建筑设计-作品集-加拿大-现代 Ⅳ.①TU206

中国版本图书馆CIP数据核字(2012)第013193号

English & Simplified Chinese Editions Are Exclusively Distributed in India by Patrika Book Centre

加拿大当代顶级建筑师

张钰华 编

责任编辑：	刘 焱 林 溪
翻　　译：	林 溪
责任监印：	彭李君
排版设计：	刘 青
出　　版：	江苏人民出版社（南京湖南路1号A楼 邮编：210009）
发　　行：	天津凤凰空间文化传媒有限公司
销售电话：	022-87893668
网　　址：	http://www.ifengspace.cn
集团地址：	凤凰出版传媒集团（南京湖南路1号A楼 邮编：210009）
经　　销：	全国新华书店
印　　刷：	深圳当纳利印刷有限公司
开　　本：	965×1270mm　1/16
印　　张：	20
字　　数：	160千字
版　　次：	2012年3月第1版
印　　次：	2012年3月第1次印刷
书　　号：	ISBN 978-7-214-07914-5
定　　价：	288.00元（USD 50.00）

（本书若有印装质量问题，请向发行公司调换）

【前言】

This book collects nearly 50 architecture projects from Canadian architects. The collected projects with different styles and unique design are the Canadian top architects' latest masterpieces. The book includes four sections: culture, commerce, residence and office. Each section has 5 to 20 projects. Every project is presented with design interpretation both in English and Chinese, real photos, plans, elevation drawings and some engineering drawings. With a high reference value, it represents the main architectural style, as well as excellent design levels of architect in the world.

This book is planned by Archiwisdom Book Studio. It serves as the valuable reference in the field of architecture practice and also one of the significant high-end readings reflecting the trends of public architecture design.

本书共收录近50个加拿大建筑作品。它们风格迥异，设计独到，为加拿大当代顶级建筑师的最新力作。本书分为四大板块：文化、商业、住宅、办公。每个板块包含5至20个最新案例。每件作品由中英文设计说明、实景照片、平面图、立面图和部分工程图组成，参考性极强，代表了加拿大的主流建筑风格以及建筑师卓越的设计水平。

本书由Archiwisdom Book工作室策划，是一本极佳的建筑实践参考读本，也是反映公共建筑设计趋势的高端读物之一。

目录
CONTENTS

CULTURE

- 8　MONTARVILLE – BOUCHER DE LA BRUÈRE PUBLIC LIBRARY
- 16　WINNERS OF THE CAHP AWARDS PAR
- 20　MONTREAL MUSEUM OF FINE ARTS
- 26　CARTIERVILLE Y CENTRE – A COMMUNITY BEACON
- 34　COMPLEXE SPORTIF DE L'ASSOMPTION
- 44　EXPANSION OF SPORTIF J.-C. MALÉPART CENTRE
- 54　PIERREFONDS COMMUNAUTAIRE CENTRE
- 62　DENISE PELLETIER – GRANDA THEATRE
- 70　GARDINER MUSEUM OF CERAMIC ART
- 80　JEAN-DE-BRÉBEUF COLLEGE – PUBLIC ART
- 84　ENTRY FOR THE COMPETITION FOR THE SAINT-LAURENT LIBRARY
- 90　EXPOSING THE INTANGIBLE: GOING BEYOND WHITE BLINDNESS
- 94　UQAM CAMPUS
- 104　DURHAM CONSOLIDATED COURTHOUSE
- 116　ROY MCMURTRY YOUTH CENTRE
- 122　JOHN MOLSON SCHOOL OF BUSINESS
- 126　MCGILL UNIVERSITY LIFE SCIENCES COMPLEX
- 136　D'ÉTUDES NORDIQUES COMMUNITY SCIENCE CENTRE

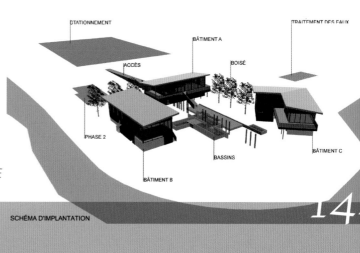

COMMERCE

- 146　LUXURY, SERENITY AND INNOVATION: LE GERMAIN CALGARY HOTEL
- 152　STRÖM SPA NORDIQUE
- 164　BRANDON NO.1 FIREHALL
- 170　LES LOFTS REDPATH
- 178　MANITOBA HYDRO PLACE
- 184　CIDRERIE LA FACE CACHÉE DE LA POMME
- 190　REFURBISHMENT AND EXTENSION OF THE CRÉVHSL HEAD QUARTERS
- 200　STATION BLÜ

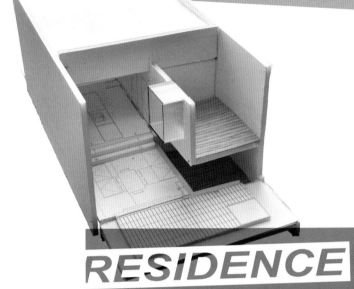

RESIDENCE
208

- 210 BERNIER – THIBAULT HOUSE
- 218 LA CORNETTE
- 226 U HOUSE
- 232 HOULE-THIBAULT RESIDENCE
- 236 ST-HUBERT RESIDENCE
- 242 CHRISTIE BEACH RESIDENCE
- 248 NIAGARA RIVERHOUSE
- 256 GEOMETRY IN BLACK
- 262 LA SOURCE – MASSAGE THERAPY PAVILION
- 268 SANCTUAIRE MONT CATHÉDRALE

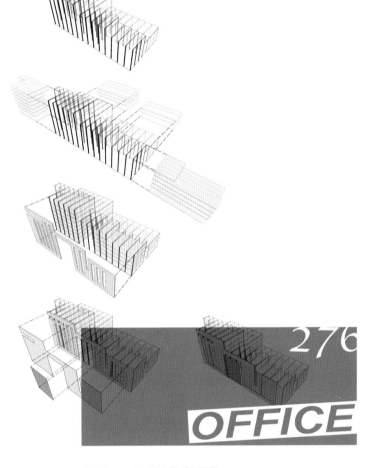

OFFICE
276

- 278 SUGAR CUBE
- 284 HEAD OFFICE OF QUEBECOR
- 292 BAY ADELAIDE CENTRE TOWER
- 300 NOVA SCOTIA POWER INC.
- 308 SIÈGE SOCIAL DE SCHLÜTER-SYSTEMS INC.
- 316 780 BREWSTER

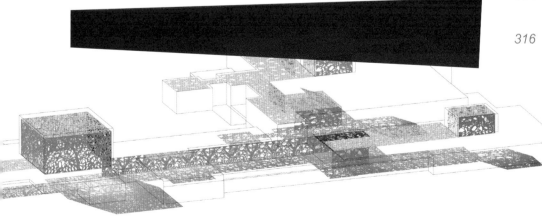

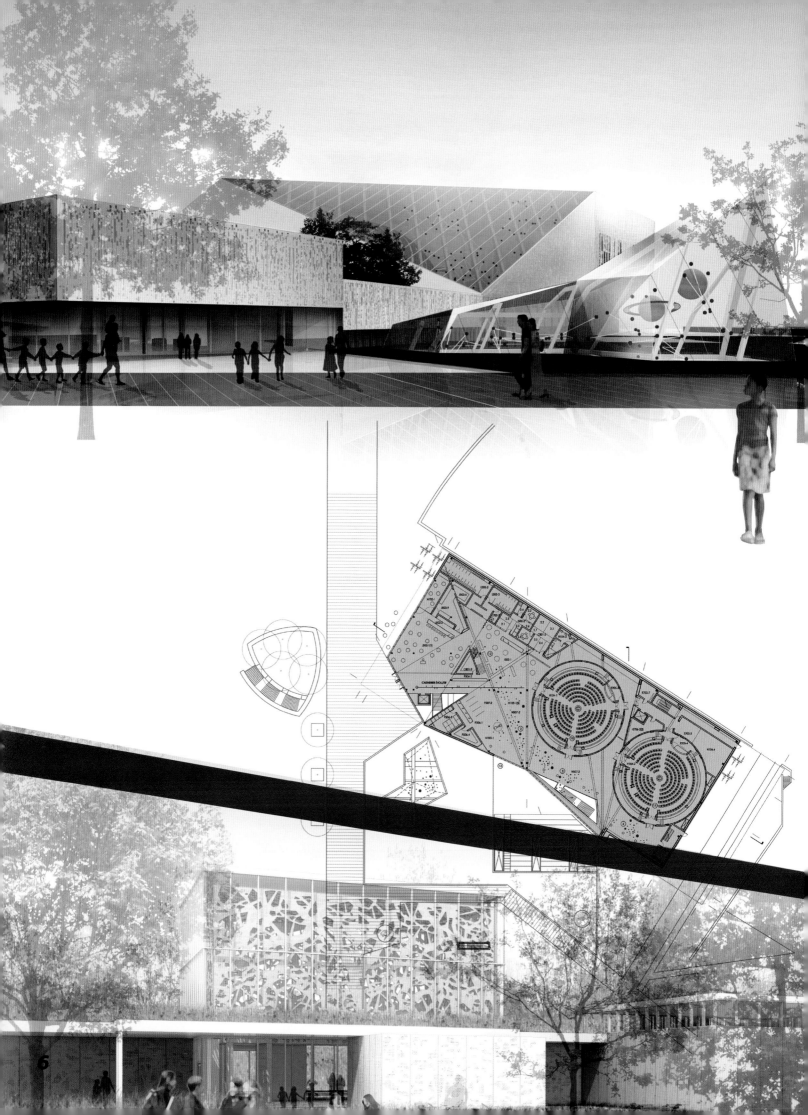

CULTURE

8	MONTARVILLE — BOUCHER DE LA BRUÈRE PUBLIC LIBRARY
16	WINNERS OF THE CAHP AWARDS PAR
20	MONTREAL MUSEUM OF FINE ARTS
26	CARTIERVILLE Y CENTRE—A COMMUNITY BEACON
34	COMPLEXE SPORTIF DE L'ASSOMPTION
44	EXPANSION OF SPORTIF J.-C. MALÉPART CENTRE
54	PIERREFONDS COMMUNAUTAIRE CENTRE
62	DENISE PELLETIER — GRANADA THEATRE
70	GARDINER MUSEUM OF CERAMIC ART
80	JEAN-DE-BRÉBEUF COLLEGE — PUBLIC ART
84	ENTRY FOR THE COMPETITION FOR THE SAINT-LAURENT LIBRARY
90	EXPOSING THE INTANGIBLE: GOING BEYOND WHITE BLINDNESS
94	UQAM CAMPUS
104	DURHAM CONSOLIDATED COURTHOUSE
116	ROY MCMURTRY YOUTH CENTRE
122	JOHN MOLSON SCHOOL OF BUSINESS
126	MCGILL UNIVERSITY LIFE SCIENCES COMPLEX
136	D'ÉTUDES NORDIQUES COMMUNITY SCIENCE CENTRE

Architects: Briere, Gilbert + Associates, Architecture & Design Urbain
Location: Boucherville, Quebec, Canada
Photographer: Christian Perreault

MONTARVILLE — BOUCHER DE LA BRUÈRE PUBLIC LIBRARY

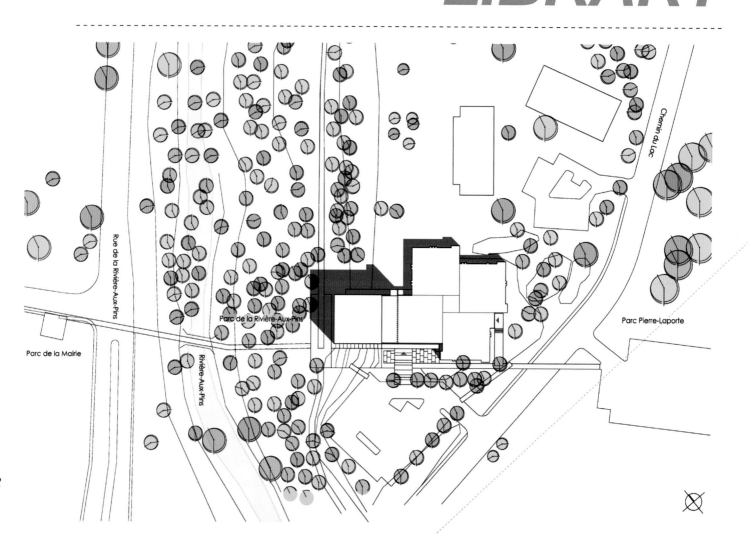

Montarville - Boucher de la Bruère Public Library is located in the downtown core of Boucherville, situated on the banks of the St. Lawrence River just east of the Island of Montreal. With a long history of almost 30 years, the municipal library needs to expand and reconfigure its existing facilities so that it can better pursue its mission and provide services in accordance with new and emerging social, cultural and technological trends.

This project, which was the award winner in a 2007 architectural competition, consists of a three-storey expansion (1,470 m^2) plus a refit of the existing structure (1,700 m^2). It includes an atrium, a new entrance hall, a new library promenade, a new loans counter and a complete reorganization of all the library collections.

In contrast to the existing building, whose introverted geometry suggests only the slightest relationship with its immediate social and natural environment, the approach adopts an open,

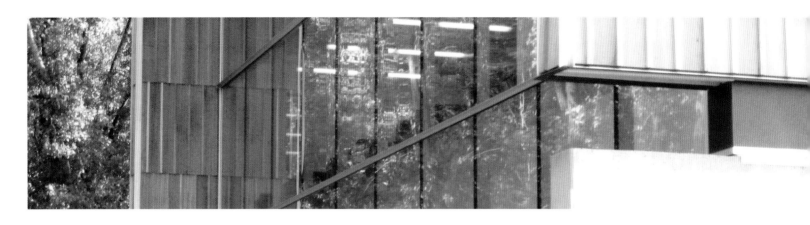

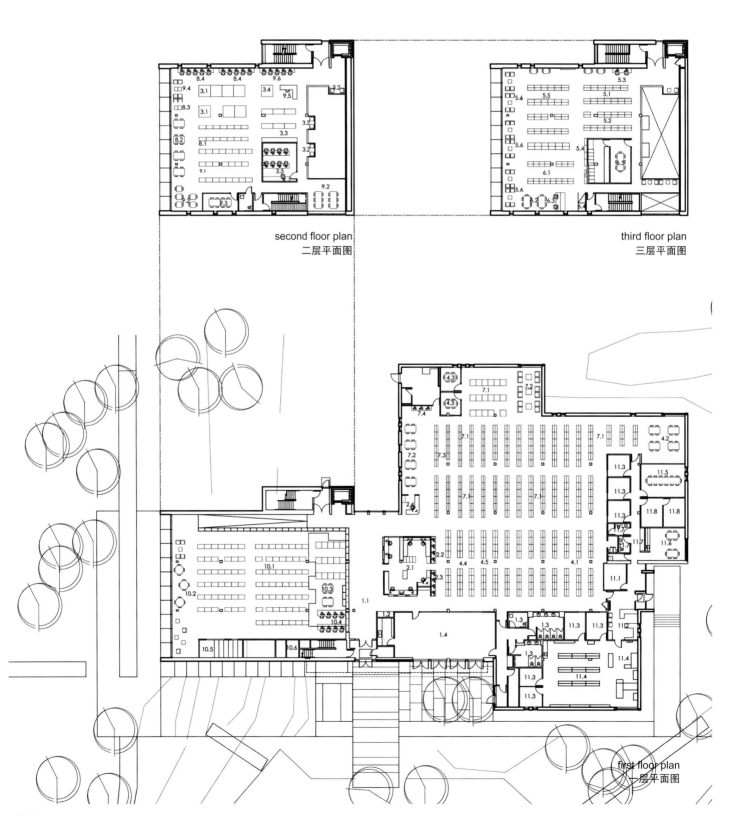

second floor plan
二层平面图

third floor plan
三层平面图

first floor plan
一层平面图

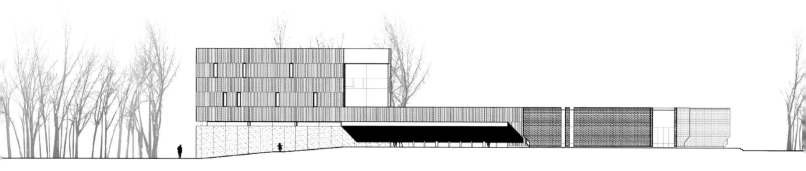

south elevation
南立面图

barrier-free design that will convey the very essence of a centre whose essential function is discovery, as well as openness to knowledge and to the world.

The expansion is simple, open and effective, to the benefit of all library patrons. Like a unifying link, it pulls together the component parts, giving concrete expression to the new physical and visual elements that connect the library to its urban context and the Rivière aux Pins Park. A new promenade serves as the path to the Rivière aux Pins Park,

longitudinal section
纵剖面图

enabling the library to be bathed in the nature, which is also the stimulus for the conceptual approach and further development of that idea.

The three storeys are home to the library's three general collections – books for children, adolescents and adults. Taking advantage of the natural topography of the site and of the proximity of the trees, a large three-storey glass wall allows for diverse visual links between the indoor spaces and the woods. Consequently, each clientele

(children, adolescents, adults and senior citizens) benefits from a distinct relationship with the vegetation, the trees and the foliage, which inspire calm, silence and rejuvenation.

Directly linked to existing footpaths, the new promenade runs alongside the building. It follows the contours of the topography and directs visitors toward the new reception area and main entrance, thereby anchoring the library to its immediate context, its neighbourhood and its town.

Montarville - Boucher de la Bruère公共图书馆坐落于Boucherville市中心，就在蒙特利尔以东St. Lawrence河沿岸。该图书馆拥有近30年的历史，这次翻修，需要对其进行扩建并更新设施，以适应日益发展的社会文化和技术潮流。

在2007年举行的建筑设计竞赛中，该设计脱颖而出并荣获大奖，包括3层楼的扩建（1470平方米），以及对现有建筑结构（1700平方米）的创新型改革。具体而言，新建了中庭、入口大厅、漫步道、借书柜台，并对

现有图书进行了整理和重组。

原先的"内向型"几何建筑构造限制了图书馆与外界自然和社会环境的沟通。与此相反，建筑师这次特意采用"开放式、无障碍"设计，将图书馆的核心功能定义为：传播知识，发现世界。

这次扩建的3层楼恰好与图书馆户外的森林公园连为一体。一条幽静的漫步小径通向美丽的Rivière aux Pins公园，使图书馆沐浴在自然之中。这也是翻修的初衷和动力之一。

这3层楼分别用来陈列幼儿图书、青少年图书和供成年人阅读的大众书籍。其高大而明亮的玻璃墙将户外的森林、绿树引入室内，成为连接室内、室外的"三棱镜"，宁静、安详、充满活力。站在这里，绿色的自然风光顿时令人心旷神怡。

漫步小径沿图书馆而建，正好顺应了其周边的地形条件，引导市民走向接待区和入口处，成为开启图书馆之旅的"引航标"，美丽、幽静、蜿蜒曲折。

Architect: Fournier Gersovitz Moss & Associates Architects
Location: Montreal, Quebec, Canada; Ottawa, Canada

WINNERS OF THE CAHP AWARDS PAR

Fournier Gersovitz Moss & Associates Architects (FGMAA) is proud to announce that it has received two awards for preservation of a heritage building under the most recent Canadian Association of Heritage Professionals (CAHP) Awards Program. FGMAA received an award of merit for the restoration of the F.A.C.E School's auditorium in Montreal, and an award for restoration of the West Block Building and Southeast Tower on Parliament Hill in Ottawa.

F.A.C.E School's auditorium in Montreal

Located in a Neo-Classical building constructed in 1914, F.A.C.E (Formation Artistique au Coeur de l'Éducation) is a public, bilingual nursery, primary, and secondary school. It was founded in 1975 to offer an alternative education program combining

最近"加拿大杰出建筑师"的系列奖项（CAHP）中，FGMAA建筑设计事务所在其中的两项榜上有名——蒙特利尔F.A.C.E学校礼堂以及渥太华国会山庄西区大楼和东南区大厦。

蒙特利尔F.A.C.E学校礼堂

成立于1975年的F.A.C.E学校，是一所双语公立托儿所和中小学。除了教授传统的文化课程，该学校还开设多种丰富多彩的艺术课程，并且，这也逐渐成为其办学的一大特色。它的教学楼就是一栋始建于1914年的"新古典主义"建筑。这一次，FGMAA建筑设计事务所应F.A.C.E学校董事会的邀请，翻修学校的会议礼堂，并将其重新定义为各种课程研发、创新之汇集地。翻修的目的在于保留其原有特色的同时，将其陈旧的设备加以更新和改造，使其焕然一新。

conventional studies and art curricula. FGMAA was mandated by the Montreal school board to restore the school's auditorium and redefine its use in the context of a changing academic curriculum. The purpose of rehabilitation of the auditorium was to both restore its original splendor and update its equipment.

West Block Building and Southeast Tower on Parliament Hill in Ottawa

In 1995, FGMAA, in association with the Arcop Group of Montreal, received the mandate to rehabilitate the West Block Building on Parliament Hill in Ottawa. The project includes the complete rehabilitation of the building and the construction of an interim House of Commons in view of the renovation of the Centre Block. Public Works Canada decided to proceed with consolidation of the Southeast Tower of the building, and to make a benchmark project for continuation of the restoration of the building's exterior. Because of the serious degradation of the tower's walls, implementation of the project necessitated the design and erection of an independent freestanding scaffolding system.

The team of architects at FGMAA proved their professionalism, efficiency, and vast experience in preservation of heritage buildings by completing these two projects with excellence and within the planned deadlines and budgets.

渥太华国会山庄西区大楼和东南区大厦

1995年，FGMAA建筑设计事务所和蒙特利尔Arcop设计院共同接到了翻修渥太华国会山庄西区大楼的任务，重新装修国会山庄，并在其内部新建一个能够直观中心街区的下议院。除此之外，加拿大公共工程部决定在此基础之上再接再厉——重新加固东南区大厦，重新装潢外表皮，使其成为该地区的一座标志性灯塔。由于该建筑的墙面材料已经老化，所以，安装一个独立非附属的脚手架系统就成为必须。

FGMAA建筑设计事务所的建筑师以其新颖的设计理念、精湛的技术工艺，以及高效的工作作风，在保留原有建筑特色的同时，在交付设计的最后期限之前，在有限的预算范围之内，出色地完成了上述两项工程。

Architects: Provencher Roy, Associates Architects
Location: Montreal, Quebec, Canada

MONTREAL MUSEUM OF FINE ARTS

Montreal Museum of Fine Arts, which celebrated its 150th anniversary in 2010, is expanding, adding the fourth pavilion on the base of the three, according to the the decisions of the board of directors. The addition of this fourth pavilion will more than double the area devoted to Quebec and Canadian art. The museum will soon be presenting a unique and coherent look at the history of Quebec and Canadian art. With admission and audioguides free of charge at all times, it will give thousands of visitors, school groups, families and tourists an opportunity to learn more about the heritage, which will be shown here in a historical context.

450 professionals and craftsmen convene on this huge building to combine the museum function with a church built in the late nineteenth century. Not only is the museum's function an ideal fit with the church, giving it a new life, it also makes it possible to conserve a "Canadian architectural treasure", with plenty of comporary fragrance. Its exceptional integration of conventional and urban design, brings past and future together. What's more, the Bourgie Concert Hall into the museum will present numerous concerts and activities every year, helping spark a new, perfect dialogue between the visual arts and music.

2010年，蒙特利尔艺术博物馆在迎来其150周年纪念之际，董事会一致决定将其进一步扩建，在原有3个庭院的基础上，增加第4个庭院，使其占地面积增加一倍。这第4个庭院将被用来陈列历史文物，同时模仿历史遗迹的建造方式，旨在为到此的旅游者、学校访问团等受众群提供一个"在历史中学习历史"的绝佳机会。

45名建筑师和工匠在此集结，秉承19世纪晚期的建筑风格，试图将新庭院打造成一个"古老的教堂"的形状。在保留"加拿大传统建筑风格"的同时，为其注入一丝丝浓郁的当代气息，亦古亦今。"教堂"内部的音乐厅，每年将举行数以千次的古典音乐会，形成视觉艺术和音乐之间完美的对话。

Architect: Daoust Lestage Inc. Architecture Design Urbain

Engineer: Dessau Ingénierie Inc.

Location: Montreal, Quebec, Canada

Site Area: 11, 853 m²

Built Area: 8,175m²

Photographers: Marc Carmer, Daoust Lestage (Marie-Josée Gagnon)

CARTIERVILLE Y CENTRE — A COMMUNITY BEACON

The Cartlerville Y Centre is a pavilion glowing through a screen of vegetation. Located in an economically depressed neighborhood; the building becomes a beacon in the urban landscape. By day, the white concrete block facade is a festive marker on a park-like site. At night, the signature lighting and large glazed openings invite the community to take part in civic and athletic activities.

The building splits into two volumes each connected to distinctly oriented North and South urban fabric, thus creating a delaminated central spine. The North and South wings become two level, inhabited sports plateaus housing the fitness centre, swimming pool, track and multi-functional rooms, respectively.

The entrance lobby, and main circulation space, is the connector of the north-south volumes, forming an interior pedestrian street. The entrance lobby and interior space are connected and split by a series of French windows, creating a strong sense of transparency and visual continuity from floor to floor and indoor to outdoor.

　　Cartierville Y中心是一片绿地之中的凉亭，位于该城市周边的经济欠发达地区，但它正逐渐成为该地区的一座灯塔。白天，其白色的混凝土建筑立面彰显着欢乐与喜庆，成为其周边公园环境中的一景。夜晚，其闪烁的灯光和敞开的大门邀请着社区居民来此参加丰富多彩的健身、娱乐活动。

　　该建筑被割裂成两个部分，分别面朝城市的北边和南边，形成一个分层式的"螺旋"。在"螺旋"的中间，分别修建了健身房、游泳池、运动跑道，以及多功能房间。

　　主厅作为主要流通空间，是南北分区的连接，在其内部，形成了一条贯穿南北的人行道。主厅和内部空间均以明亮的落地窗相互区隔，站在这里，就能将Cartierville Y中心的全部布局尽收眼底，在视野上一气呵成。

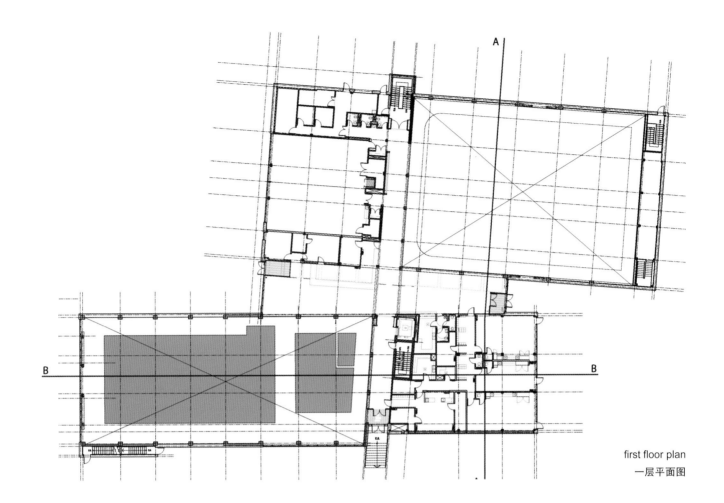

first floor plan
一层平面图

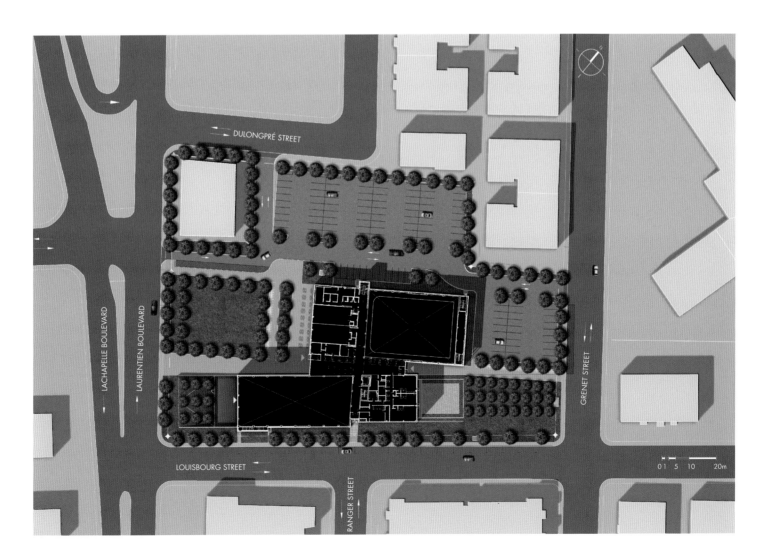

On the lower level of the southern volume, the pool deck extends West to the level of the street. To the north, the track surrounding the upper perimeter of the gym projects the runner into nature.

From the urban gestures to the architectural resolution, the Cartlerville Y Centre is designed with a holistic approach to sustainability. In combination with the white facade and white roofing membrane the heat island effect is reduced and a geothermal system completes the sustainable strategy.

The architecture also strives to foster sustainable communities. The transparent facade invites the public to become a part of the community and of the centre. On the upper level, the fitness centre, swimming pool, track and multi-functional rooms are arranged on either side of the building's spine. These spaces benefit from open views to each other as well as elevated views towards the lobby, pool and gym below showcasing community life. As a result of the generous transparency of the spaces, both inwards and outwards, the Cartlerville Y Centre projects a genuine desire for openness to the community.

在南面分区的底层，泳池甲板向西延伸，一直通向户外的街区。北面，围绕健身房而建的小型跑道亲近自然，同时在阳光的照射下，运动者与其自己的影子相映成趣。

与城市可持续发展理念相一致，该中心白色的立面和屋顶薄膜有效地抑制了城市热岛效应。其内部独立的地热系统，实现了可持续性的能源供应。

Cartierville Y中心的设计旨在号召丰富、和谐的社区生活。透明的整体外观邀请社区居民前来成为其中的一员。健身房、游泳池、幽静的运动跑道，以及多功能房间，蜿蜒地盘旋于"螺旋"之上。得益于开阔的视野，这些设施彼此相连，浑然一体。

Cartierville Y中心已经成为该城市中一道亮丽的风景线，彰显着开放、多元化的社区生活特色。

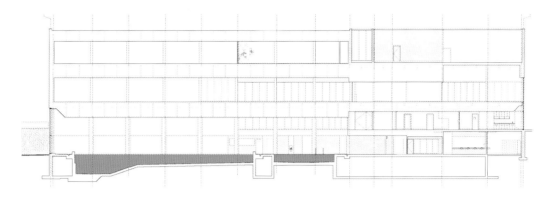

section BB
剖面图BB

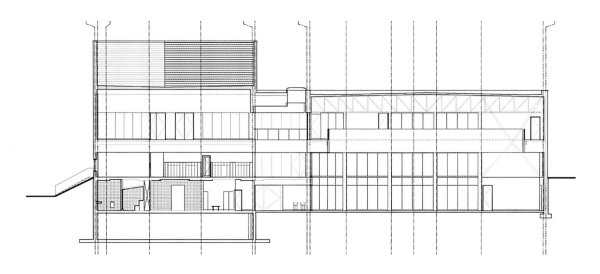

section AA
剖面图 AA

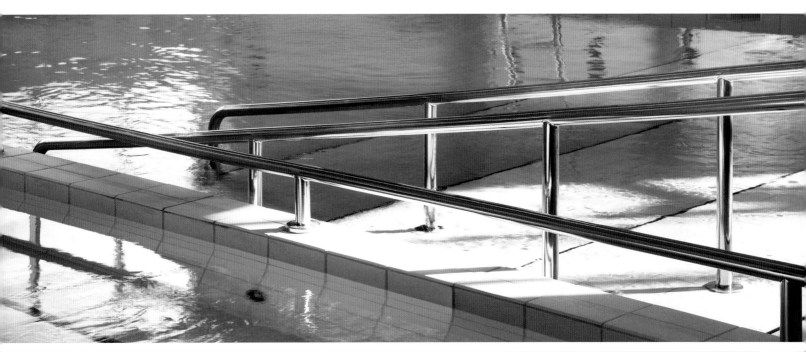

Architect: Les Architects FABG

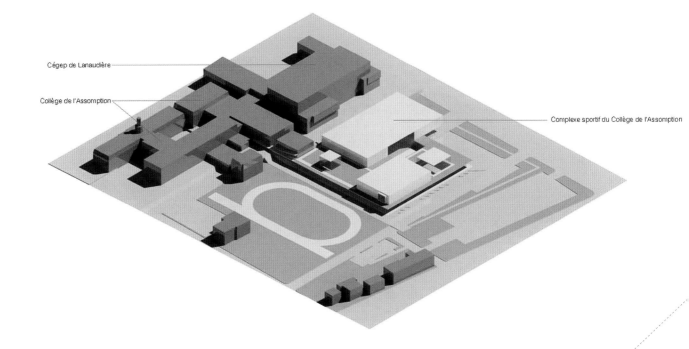

COMPLEXE SPORTIF DE L'ASSOMPTION

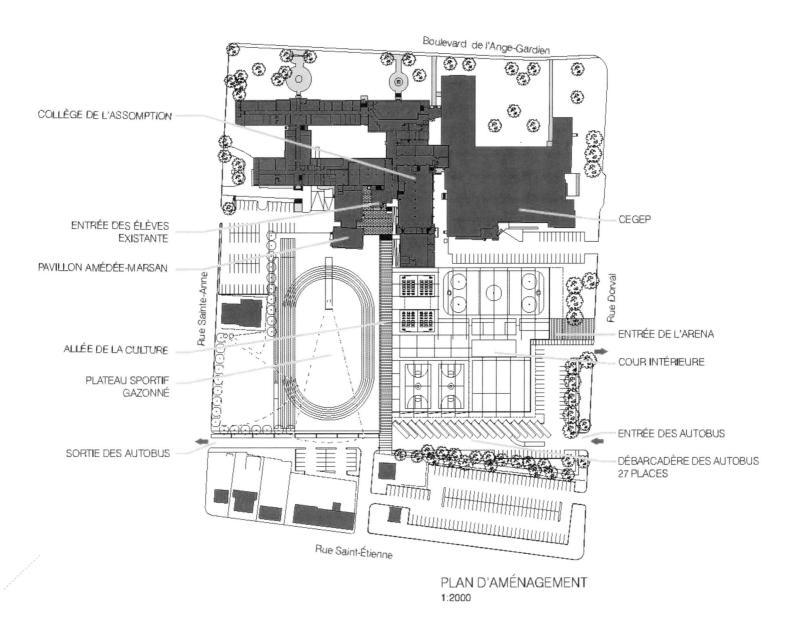

PLAN D'AMÉNAGEMENT
1:2000

L'Assomption College is a private high school founded in 1884 in a small town near Montreal. This college, that hosts 1,250 students, has become the heart of the social, cultural and sportive life of the Lanaudière region. The project aims at efficiently waking use of college space and providing functional infrastructure.

The crumbling municipal hockey arena that was also built alongside the college needed to be replaced. It led to the decision to build a new sport center including a double gymnasium, workout and dance rehearsal rooms and a new ice rink.

The existing hockey arena was demolished to reorganize the site, clearing a large space for a new football field and race track along the sport centre. It also cleared views on the existing Amédée-

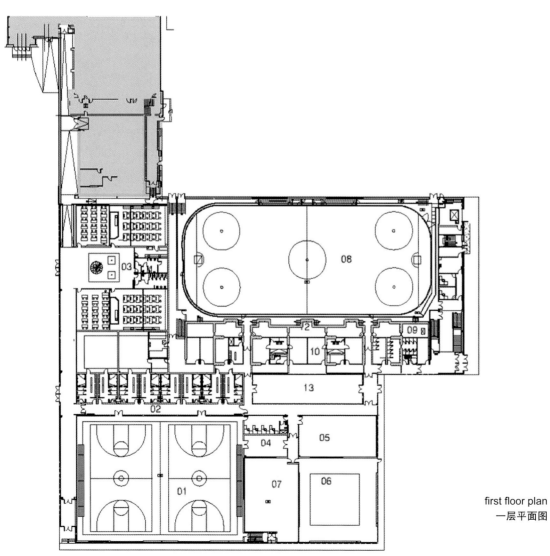

first floor plan
一层平面图

36

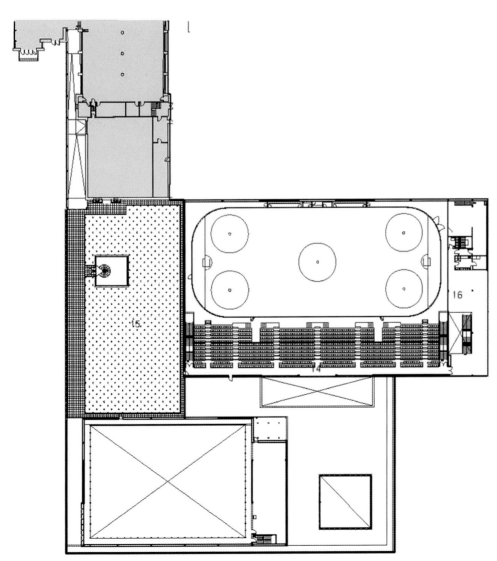

second floor plan
二层平面图

37

Marsan pavillion, gave a distinctive entrance to the new arena allowing that it operates independently and helped reorganize the students arrivals and departures area that was originally chaotic.

The formal strategy used in this project takes advantage of the level difference between the ground floor of the existing college and the construction site to establish a continuous horizontal datum at mid height between these two levels all along the perimeter of the new building. The lower half is largely glassed while the volumes of the arena and gymnasium, the glass box of the green roof access and the dance rehearsal room emerge above this base. The roof which is at the same level as the ground floor of the existing college gives a privileged point of view to watch events taking place on the sport field.

　　L'Assomption学院是位于蒙特利尔附近郊区的一所私立高中，成立于1984年，拥有1 250名学生，成为该地区社会、文化和体育活动的中心。此次翻修，旨在更加合理高效地利用校园空间，提供更加实用的基础设施。

　　经过缜密的考量，建筑师决定将该学院校园沿线的曲棍球运动场拆除，在此建造一个综合性的体育馆，包括运动操场、两个健身房、舞蹈声乐室、溜冰场，这样，校园空间被更加高效地利用起来。

　　曲棍球运动场的拆除，将原本就很狭窄的校园沿线解放出来，又在此建造了一个新的足球场和几条跑道。站在这里，能够直接眺望到远处的Amedee-Marsan凉亭，同时解放出来的空间也有利于在每天上学放学的时候，有秩序地组织学生队伍，消除了原来的混乱。

　　建筑师利用教学楼和体育馆之间的高度差，在它们之间建造了一个水平的基准面。运动操场、健身房、舞蹈声乐室都在它之上。站在与教学楼一层平行的露台上，能够将整个体育馆的美景尽收眼底。阵阵清风吹来，美不胜收！

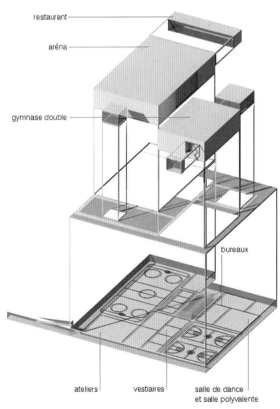

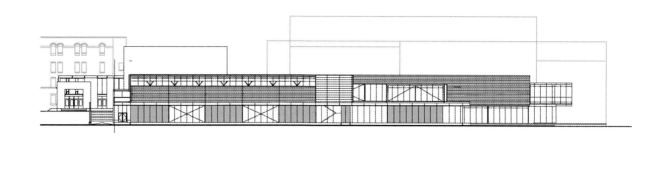
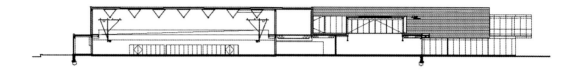
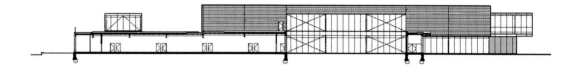
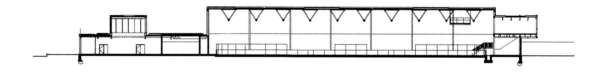

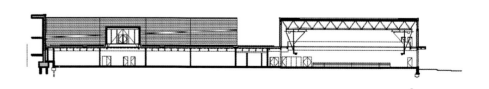

41

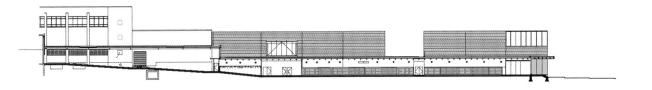

Architect: Saia Barbarese Topouzanov Architects

Photographers: Marc Cramer, Frédéric Saia, Vladimir Topouzanov

EXPANSION OF SPORTIF J.-C. MALÉPART CENTRE

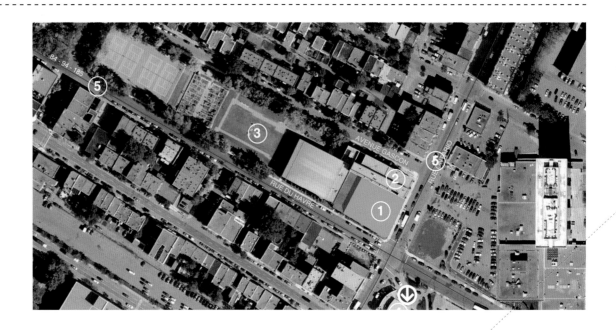

The new building had to complement the surrounding urban fabric and form a unit with the articulated volumes of the 1996 sports centre.

The project has creatively applied the idea of waves. Both indoors and outdoors, a wave develops on the perimeter of the building, rising and falling in two continuous undulations. The first, transparent and high opposite to the entrance and the training room, surges as far as the corner near the diving board, providing an ambience propitious to concentration. Then the glass rises along the wading pool and reflects the green world outside, conveying the idea of "healthy sports"

The dialogue between the new building and the old one is natural The grey-tinted glass harmonizes with the colors of the earlier structure. It gives passers-by an idea of what is going on inside while preserving some privacy for users. The milky tone and reflective quality of the upper wall are echoed in the material for the roof to ensure the extension from one to the other in contrast with the clear demarcations of the building.

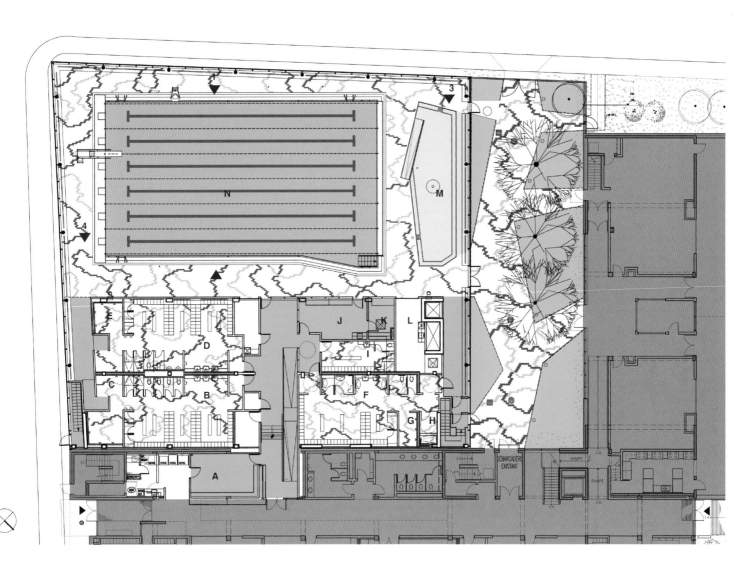

first floor plan
一层平面图

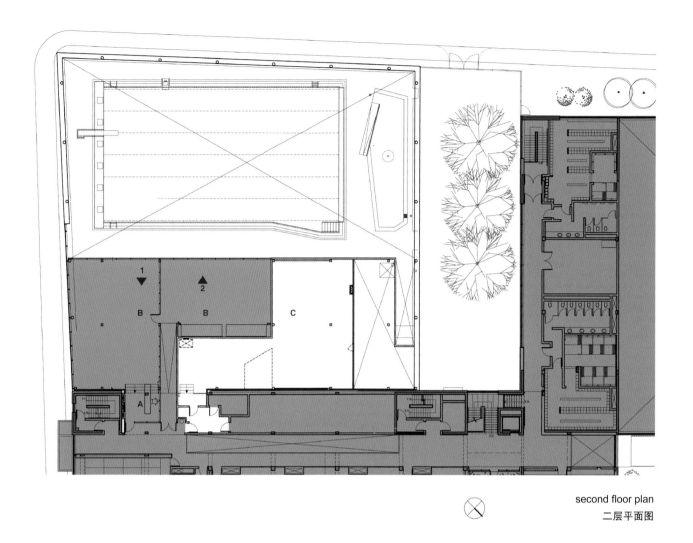

second floor plan
二层平面图

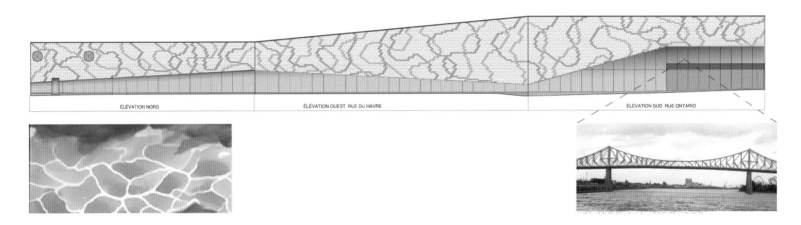

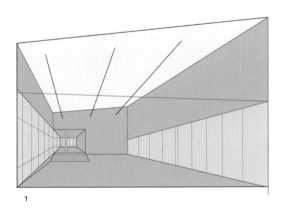

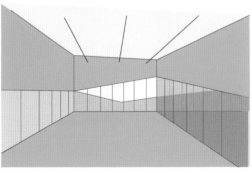

This architectural project is, above all, people-oriented. It is respectful of passers-by, who find themselves walking beside a shimmering wall beyond which they can divine movement inside, and the entrance beckons like an invitation. The project is even more concerned with the well-being and health of users, encouraging passers-by to develop healthy lifestyle. The construction process, the systems implemented, and the air and water recycling were all designed with sustainable development in mind. All of these concerns modulate an architectural language in step with current ideas, locally and internationally, that the pursuit of the healthy, high-quality life has already come into being, not a dream.

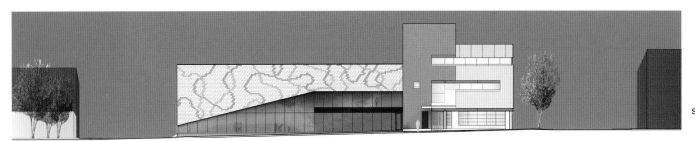

south elevation
南立面图

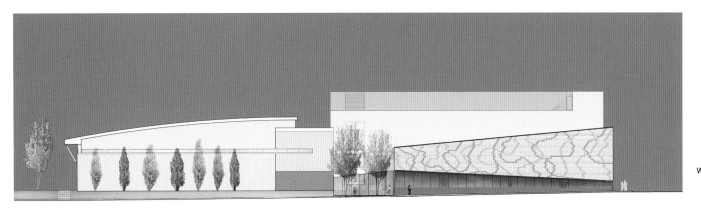

west elevation
西立面图

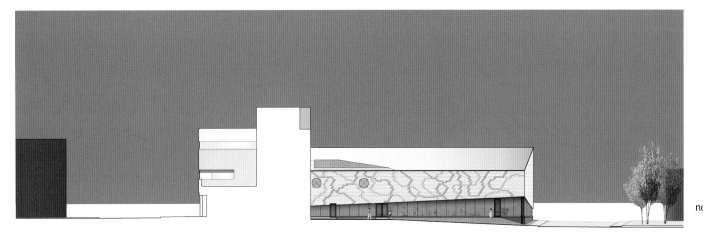

north elevation
北立面图

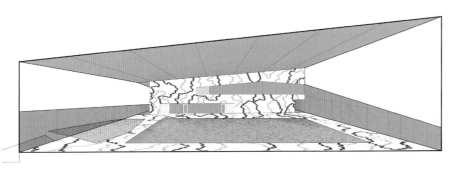

1

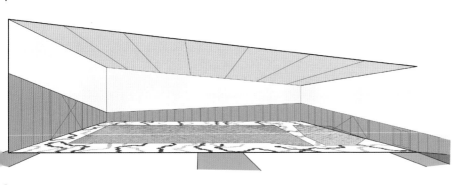

2

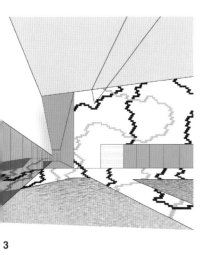

3

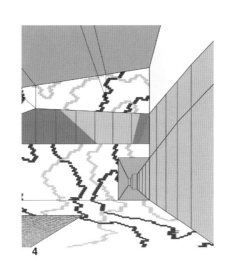

4

该项目是对1996年建成的Sportif J.-C. Malépart体育馆的延伸，将旧建筑周边的空闲土地加以合理地利用与开发，旨在在其旁边建造一个新型体育健身场所。

该项目创造性地采用"波浪式新型设计理念"。在室内和室外，在建筑的周边，一条巨大的波浪，此起彼伏：一方面，入口处和训练中心对面的波浪式跳水板，有利于提高空间利用率；另一方面，笼罩在游泳池上方的波浪式玻璃，反射着远处花园的绿意盎然，更与点点绿意相映成趣。

新旧建筑之间形成自然、和谐的对话，全然没有新旧之间生硬的隔阂。茶色隔热玻璃与旧建筑的颜色一致，使新建筑保持了良好的私密性。路过此处，使人好奇，里面的世界究竟是怎样的？奶茶色墙面和屋顶相得益彰，同时又令该建筑的各个部分具有清晰的界限，各显特色。

该建筑以"以人为本"为核心目标，号召人们注重健康。它仿佛是一张"做运动的邀请函"，邀请路人在此驻足，在此养成健康的生活习惯。施工过程中的空气和水资源的循环式设计都体现了可持续发展的基本原则。这里所有的建筑语汇都证明：对于健康、高质量生活的追求，已不再是不切实际的空想，它已经成为实实在在的实践。就在此时，就在此处！

5

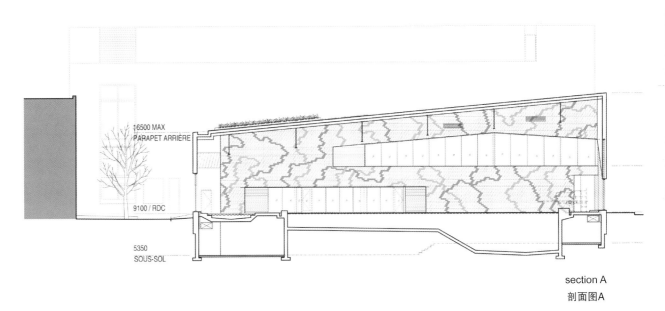

section A
剖面图A

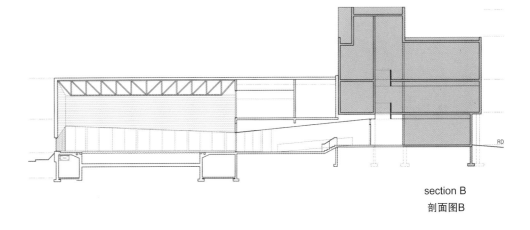

section B
剖面图B

Architect: Les Architects FABG

PIERREFONDS COMMUNAUTAIRE CENTRE

The Pierrefonds-Roxboro borough located on the west end of Montréal's island contains a population of more than 65,000 inhabitants. It has an excellent location. The proximity of a train station, a school, a youth centre and a park led the municipal authorities to implement a new community center facilitating the integration of new citizens with the help of social organizations and volunteers.

To avoid institutional stigmatization, the project adopts the morphology of commercial buildings: one-storey building with large fenestration and a overhanging roof on small columns.

This voluntary simplicity enabled us to provide the project with a green roof, an outdoor bandstand. The green roof serves as a perfect protection from the heavy sunshine in summer and standing on the bandstand, it is pleasant to sing a song.

A cladding of silicon coated glass panels and a glossy powder coating on aluminium plates were chosen for their resistance to graffiti, also ensuring that it is full of natural country fragrance.

The centre includes a multipurpose room for 150 persons, classrooms of different sizes and a communal kitchen, which is beautiful and highly functional.

Pierrefonds-Roxboro自治镇坐落于拥有超过6.5万居民的蒙特利尔西海岸。它的地理位置优越，临近火车站、学校、青年中心和公园。市政府决定在其内部建造一个新型社区交流中心，旨在促进当地居民的交流与融合。

为了对传统的建筑设计模型提出挑战，摒弃"墨守成规"，该设计采用多样的商业性建筑设计模型，将这里打造成一个"生动的故事"：在只有一层的建筑中，高大的窗户上悬挂精致、独特的小物件，作为装饰；绿色的屋顶成为远离夏日里太阳光暴晒的天然屏障；站在户外的音乐台上面，放声歌唱。一切美好的故事，都从这里开始。

该建筑外观使用的硅脂玻璃钢板和光滑的铝合金钢板可以有效地防止在其表皮上做涂鸦，同时，这些取材于当地的原生态材料，也使该建筑具有浓郁的乡土商业气息，朴实且自然。

该社区交流中心包括一间能够容纳150人的多功能套房、不同规模的教室，以及一间社区厨房。这一设计精湛的建筑在打造一场视觉盛宴的同时，也具有高度的实用性。

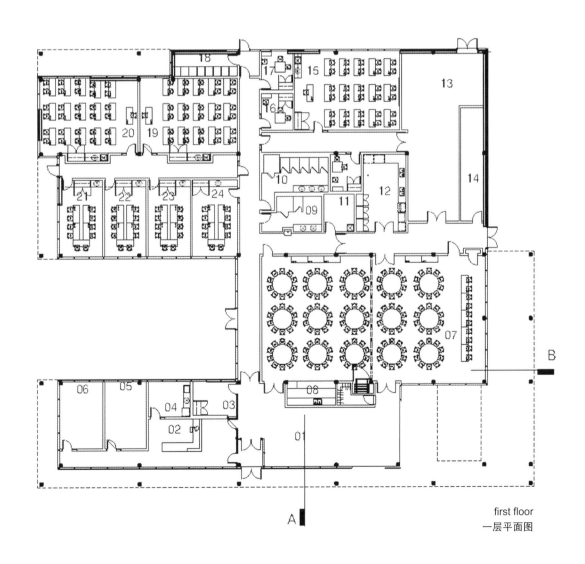

first floor
一层平面图

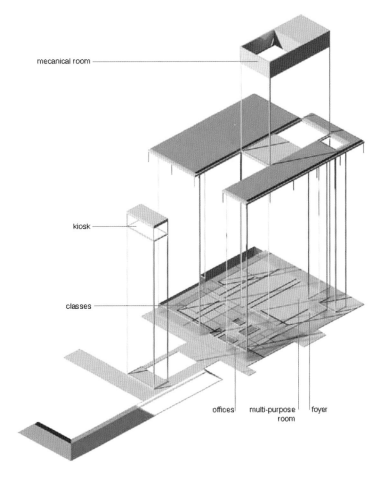

mecanical room

kiosk

classes

offices | multi-purpose room | foyer

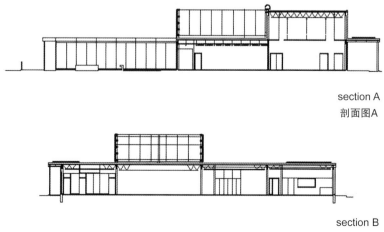

section A
剖面图A

section B
剖面图B

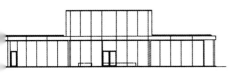

elevations
立面图

Architect: Saia Barbarese Topouzanov Architects
Area: 4,560 m²

DENISE PELLETIER — GRANADA THEATRE

Granada Theatre, built in 1930, is situated in the former municipality of Maisonneuve, at the northwest corner of Boulevard Morgan and Rue Sainte-Catherine. In the vicinity, prestigious buildings such as C. L. Dufort's City Hall, as well as the Maisonneuve Market and Maisonneuve Baths, both by M. Dufresne, testified to the city's prosperity and its architects' talents in the early twentieth century. The Beaux-Arts style then in fashion borrowed columns, arcades, and motifs from Antiquity or the Renaissance. In line with this trend, the Granada was among the wave of North American movie theatres that looked like palaces, with chandeliers, drapes, loges, gold-leaf paint, faux-marble finishes, and more.

Intending to refine a "Spanish atmospheric design," developers asked architect Emmanuel Briffa to take over the Granada project Briffa, an experienced decorator, had designed numerous theatres in eastern Canada, almost twenty of them in Montreal and Quebec City.

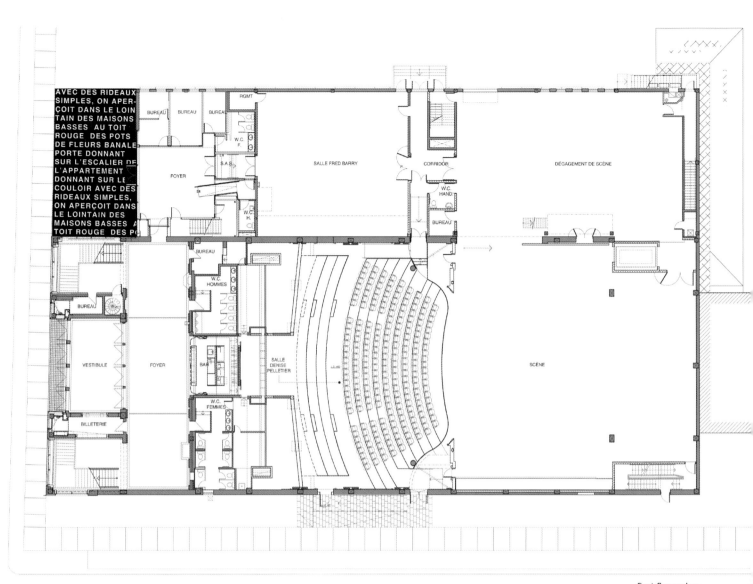

first floor plan
一层平面图

始建于1930年的Granada剧院，原先是Maisonneuve市政大楼，坐落于Morgan和Rue Sainte-Catherine林荫大道的西北角，毗邻著名的市政大楼、市场和洗浴中心，它们共同见证了城市的繁荣，以及20世纪早期建筑师扎实的设计功底以及成熟、精湛的技术和工艺。这些建筑借鉴并吸收了文艺复兴时期圆柱、拱廊、图案的建筑技艺，具有很强的古典主义美学价值。Granada剧院沿袭北美剧院的设计风格，枝形吊灯、垂挂窗帘、包厢、金色树叶喷漆、人造大理石装潢，这一切看起来好像一座富丽堂皇的宫殿。

这次整修，开发商希望将Granada剧院打造成一个洋溢着浓郁的西班牙宫殿式艺术气息的灵性剧院，并委托Emmanuel Briffa建筑师完成全部设计。他是一位杰出的建筑师，先后完成了加拿大东部众多剧院的设计，其中有20个分布在蒙特利尔和魁北克。

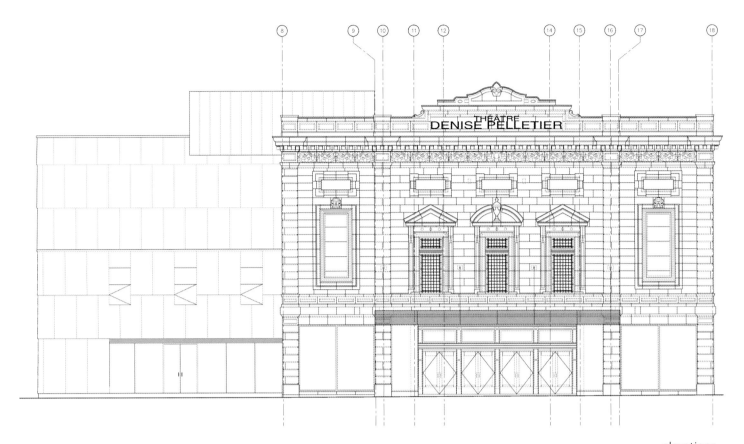

elevations
立面图

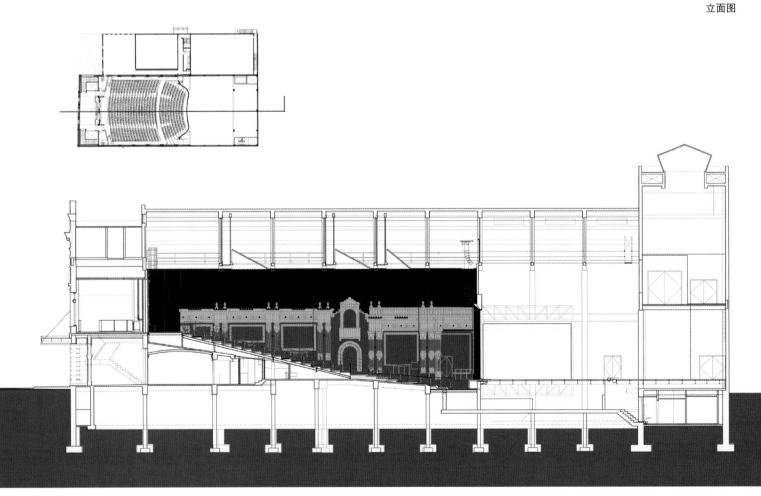

section
剖面图

In the original building, the tone was set the moment one set eyes on the theatre. The front façade had an elaborately patterned cladding of prefabricated, beige-tinted stones. The basic rectangular shape was crowned with a cornice with modillions, a frieze, and a small pediment featuring the theatre's name, "Granada." Horizontally, the façade was subdivided into two registers. The ground floor offered a wide opening with central doors sheltered by a light wrought-work marquee, flanked by display windows. The upper floor, taller than the ground floor, extended the tripartite vertical subdivision. In the centre, three Renaissance-style windows rise above the entrance. The rest of the building is clad in brick, with simple lines and easy procedures.

The luxurious interior matched the exterior. From a rather plain vestibule, visitors enter the lobby. The drop-arch moldings of the coffered ceiling sit on the brackets. The lobby features a fireplace and a fountain, like a beautiful picturesque, with a strong sense of visual impact.

Granada Theatre serves as a visual feast of luxurious design and elegant art fragrance. It deserves the classical model in the field of theatre architecture, with dynamic vibrancy and forever warmth.

该建筑规则的长方形外观上镶嵌着带有飞檐托的檐口、中楣和三角楣饰，它们形成剧院鲜明的特色。在水平方向上，建筑立面被分为两个登记注册区。在低层开阔的空间中，在窗户旁边，做工精细的天幕遮挡了中央大门。高层空间更加挑高，延伸了三角形的功能区。在中央区域中，三扇具有文艺复兴时期建筑特色的窗户从入口处缓缓升起。该建筑的其余部分被瓦砖所包装，线条简约，装修程序简便。

室内空间沿袭奢华、高雅的建筑风格。穿过一个非常平坦的前厅，观众就可以进入主厅。壁炉中燃烧着旺盛的火苗，旁边的喷泉则缓缓地倾吐出一帘帘清澈的瀑布。呈拱门形状的咖啡色天花板与壁柱上的精美图案相映成趣。这一切，成就了豪华、高雅的室内空间，好像一幅色彩艳丽的风景画，形成强烈的视觉冲击。

Granada剧院集奢华、高雅的建筑风格与浓郁的艺术气息于一身，是剧院建筑实践中的经典之作。在这里，活力激荡，温情永存。

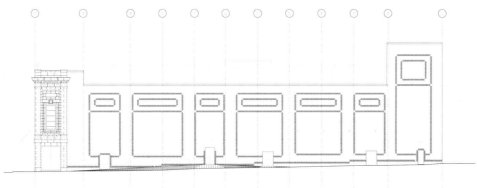

east elevation
东立面图

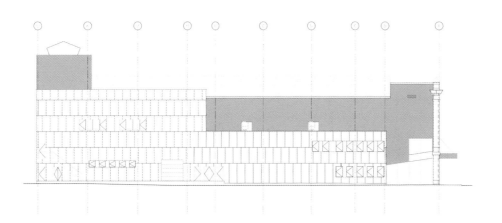

west elevation
西立面图

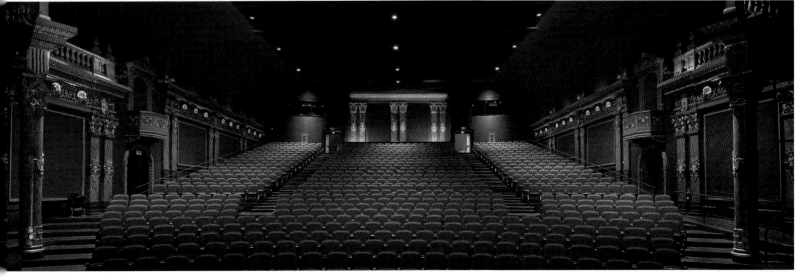

Architect: KPMB
Location: Toronto, Ontario, Canada

GARDINER MUSEUM OF CERAMIC ART

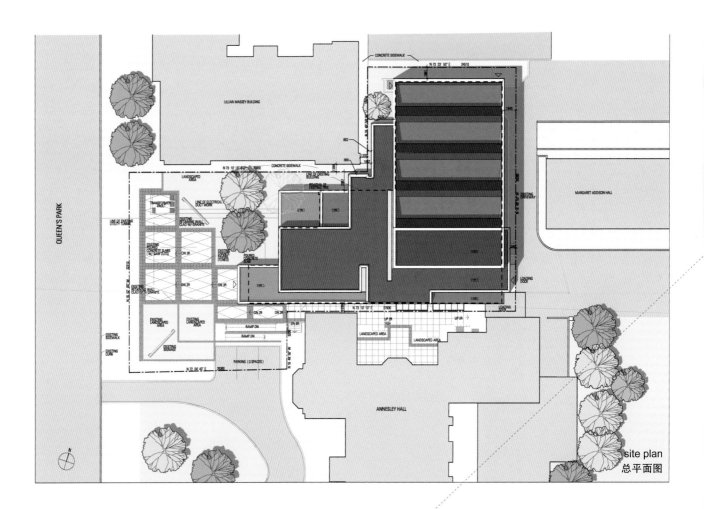

site plan
总平面图

Gardiner Museum of Ceramic Art is one of the world's pre-eminent institutions devotedto ceramic art, and the only museum of its kind in Canada. It is also one of the major projects in Toronto's cultural renaissance. The Gardiner renewal, together with the Royal Ontario Museum across the street and the Royal Conservatory of Music around the corner on Bloor Street West, will form a new cultural precinct for the city.

Framed between the neoclassical Lillian Massey building to the north and the Queen Anne-style Margaret Addison Hall to the south, it has an excellent location.

Gardiner陶瓷艺术博物馆是世界著名的艺术圣地,也是加拿大陶瓷艺术领域唯一的一个博物馆,它是多伦多许多文艺复兴项目的实施地。它与其周围的Ontario皇家博物馆和Conservatory皇家音乐厅共同形成该城市的文化基地。

该博物馆地理位置优越:北面与新古典主义风格的Lillian Massey大楼相望,南面毗邻皇家大厦。

The addition of approximately 1,300 m² creates a new contemporary gallery to host international exhibits of large-scale contemporary works, provides much-needed storage for the expanding permanent collection, and incorporates new studio and curatorial facilities to support the Gardiner's popular community-outreach programs and its research activities. The design also greatly enhances the museum's revenue-generating potential with a larger, more accessible retail shop, a rentable multi-purpose event space, and a destination restaurant.

这次翻修，旨在将Gardiner陶瓷艺术博物馆打造成一个兼收并蓄、举世瞩目的文化圣地。将在原有基础上修建一座占地1 300平方米的新馆，用来陈列大量国际著名的陶瓷艺术展品，拓宽原有的储藏用地，同时配备新摄影棚等新型馆藏设备，从而为在此举办的形式多样的社区文化和学术研究活动提供便利。这一举措将会吸纳众多的游客，从而大大提高周边餐厅和零售商店的经营效益。

The renewal builds on top of the original structure, designed by Keith Wagland in 1984, to anticipate vertical expansion. The original pink granite exterior is replaced with polished buff limestone to give the Gardiner a more contemporary image. The limestone seamlessly weaves existing and expanded spaces together. The front of the museum is completely re-landscaped with a series of terraced platforms that provide a gradual ascent into the forecourt of the building.

A cubic volume marks the entrance to the building. During the day, the cube's broad expanse of floor-to-ceiling glazing creates a reflective surface which mirrors

the outside street, and at night, acts as a window into the museum's activities. This column-free area with a clerestorey ceiling creates a monumental space for large-scale contemporary and traveling exhibits.

The retail store and restaurant are on the corner of the museum, providing exquisite souvenirsand a delicious feast.

The design emphasizes a subtle interplay between transparency and lightness, The consistent language of materials, custom-designed casework, and precise detailing provide a miracle in the field of contemporary art architecture.

新馆被建在于1984年由建筑师Keith Wagland修建完成的原有建筑之上，形成一个垂直的结构。原来的粉红色花岗岩表皮被暗黄色的石灰石所代替，使整体建筑风貌焕然一新。石灰石在新馆和旧馆之间迂回前进，将每一个角落细腻地连接起来，形成了统一、和谐的外观。建筑前方的空地被加以有效地利用，在此搭建了一系列露天阳台，呈逐渐上升状。

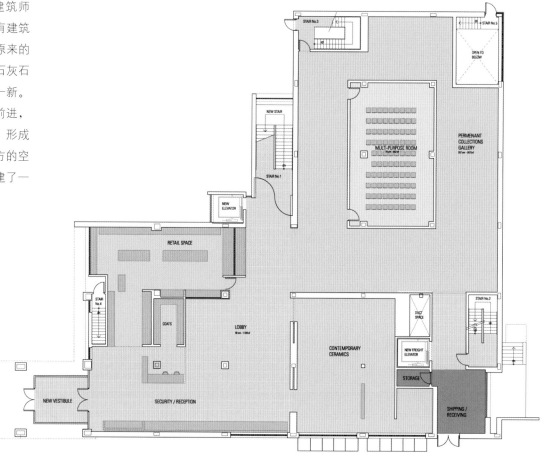

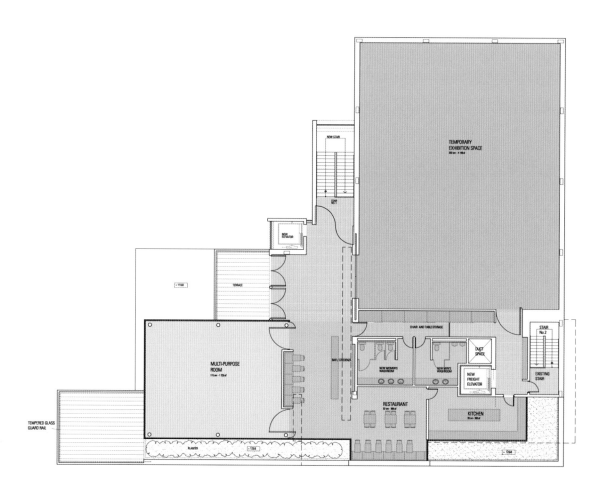

游客通过立方体式的入口进入其中，首先映入眼帘的是明亮、清澈的落地窗。白天，它将街区的自然景观完美地引入室内，站在窗前，放眼望去，令人大饱眼福；夜晚，它则将室内精美的展览品包揽其中，形成神奇的"反射世界"。高大、宽阔的天花板将馆内的艺术品笼罩其中，使这里的一切都洋溢着浓郁的艺术气息。

零售商店和餐厅处在博物馆一隅。在参观之后，游客可以进入商店挑选精致的纪念品，或在餐厅中饱餐一顿。

Gardiner陶瓷艺术博物馆建筑旨在追求透明和光亮的有机结合。连贯的建筑语汇、定制的材料、精心的细节设计，使其成为当代艺术建筑领域内的一枝奇葩。

Architects: Menkès Shooner Dagenais, LeTourneux Architects

Artist: Yechel Gagnon

Location: Montreal, Quebec, Canada

Photographers: Marc Cramer, Alexandre Masino, Pierre Charrier

JEAN-DE-BRÉBEUF COLLEGE — PUBLIC ART

The project aims to create a novel space for the Jean-De-Brébeuf College.

The space offered at the college, through the long corridor and the grand foyer, is double and unique; double in its dimensions and functions, unique by its majestic architecture that embraces the exterior garden and its natural elements. The designers thus chose to use the two different techniques that may be seen here, frottage drawing and carved plywood.

Frottage drawing serves as a kind of brand-new drawing technology created by displacement of the paper. It presents an imagery oscillating between abstraction and an imaginary landscape echoing a certain tradition of Chinese ink painting.

Carved plywood, consisting of different formats, presents various vertical architectural lines of the wall. They accentuate the rhythm in symbiosis with the architecture. Wood, an organic element, crosses the transparency of the window to rejoin the beauty of the exterior garden, which is full inspiration.

The convergence of frottage drawing, carved plywood and garden with inspirable scents, will have the pleasure to find themselves in presence of a single work of art, like a beautiful, harmonious song.

It is a marvelous college space, providing memorable enjoy.

该设计旨在为Jean-De-Brébeuf高校打造一个独树一帜的教学空间。

穿过长廊和大厅走道的教学空间，具有双重性和独特性：双重性在于其尺寸和功能，独特性在于其宏伟的建筑形式。该教学空间能够将外部的花园和自然元素包揽其中。于此，建筑师特意选择了两种不同的建筑技艺——擦印画和雕刻胶合板。

擦印画是一种全新的绘画工序，即在有位移的墙面上形成动态的图画，旨在形成中国传统水墨画与抽象绘画之间的震荡感，以及不知不觉的流动感。

雕刻胶合板呈不同的样式，位于纷繁交错的墙面线条之间，和整体建筑共同谱写着一首音调和谐的乐曲。其最重要的组成元素——木头，穿过透明的窗户，将内部空间的快乐延伸至室外的花园，丰富了花园的色彩，并且为其带来灵性，使花园不再成为"灵性设计"的奢望。

雕刻胶合板的空间构造、擦印画的动感墙面、灵性的花园，仿佛共同在该教学空间中吟咏着一首首悦耳动听的诗词。在这个别具一格的教学空间中驻足，定会让人流连忘返！

Architect: Chevalier Morales Architects FABG

ENTRY FOR THE COMPETITION FOR THE SAINT-LAURENT LIBRARY

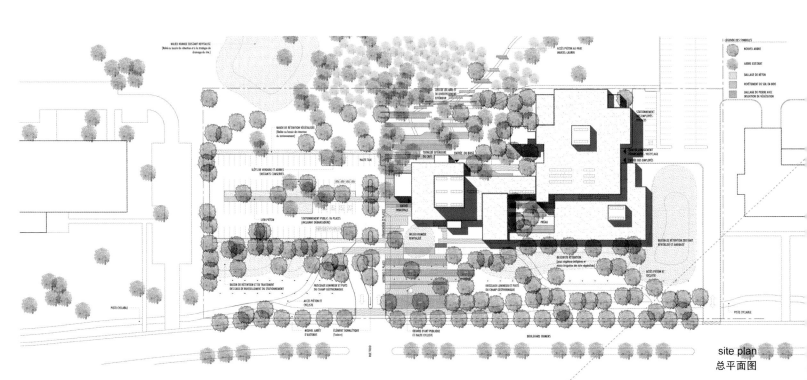

site plan
总平面图

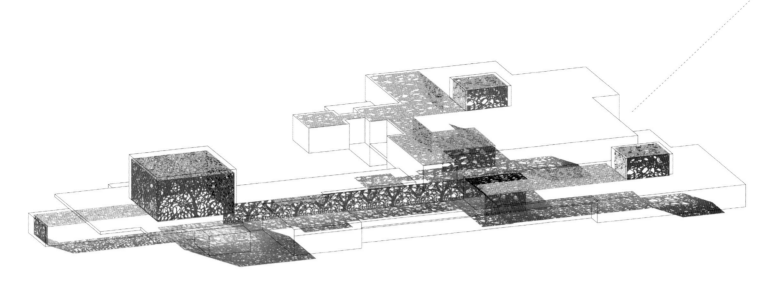

Instead of considering the new Saint-Laurent Library as "a gateway into Marcel-Laurin Park", the designers would rather imagine that it would become part of the park. In their opinion, the nature/culture dichotomy that leads to think that parks are green oasis's completely separated from the urban world by defined boundaries no longer corresponds to the aspirations of Montreal's citizens. The desire to make the city greener expands to streets, back alleys, backyards and rooftops and associates this willingness to sustainable design, quality of life and overall improvement of the city's vegetal coverage. Visiting the library can then become part of an immersive experience that contributes to this desire while offering a favourable environment for reading.

The buildings fragmented geometry allows it to blend into the existing forests, where, in particular, the cottonwoods have a life span of about twelve years. And to further contribute to its biodiversity, the disigners have proposed the planting of new varieties of trees. The library's perimeter would then be stretched to the main Boulevard so that the library could be set in the heart of the forest.

Forests provides green energy and create an air-and-noise filter. All the LED lights use solar energy. Rainwater retention surfaces are also designed to become part of the humid zones already present in Marcel-Laurin Park. Generous windows allow the user of the building to have constant visual contact with the surrounding forest. The library project offers an opportunity to create a green energy themed garden that would become a showcase of the best sustainable practices for people visiting the park recreationally or for those who walk by daily.

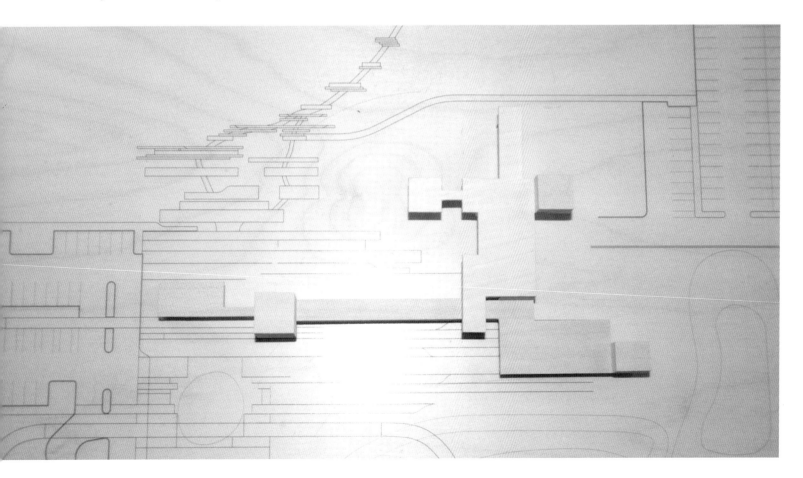

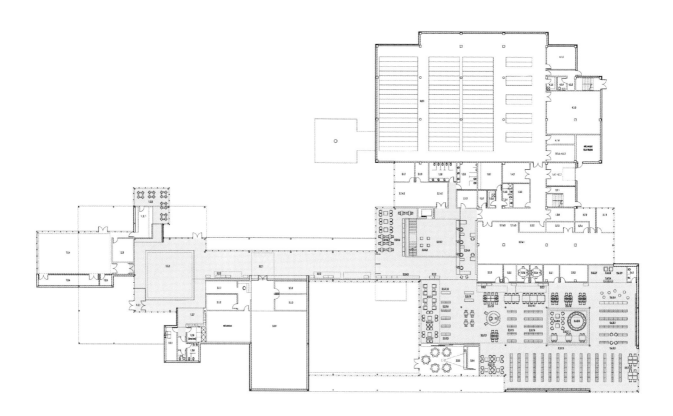

first floor plan
一层平面图

The project also includes museum storage for the city. This space is simple, rectangular and functional. Certain artefacts from the city's collections could be temporarily exposed the building's exhibition cabinets, in order to contribute making the library a lively and dynamic space.

The Periodical Room and the space used for selling books in the secondary hall also contributes making the library a public animated space, perfect for cultural exchanges while preserving a calm atmosphere associated to the delicate light, which is the key to the new library's identity.

建筑师摒弃"将图书馆作为公园大门"的传统设计模型，而另辟蹊径，使Saint-Laurent图书馆成为Marcel-Laurin公园内不可或缺的一部分。他们认为，自然和文化的分离，常常令公众认为公园是都市生活中的绿洲，"使城市绿起来"的运动，不应仅仅停留在街区、小巷、庭院、屋顶，而是要渗入都市的每一个角落。基于这种理念，建筑师致力于将Saint-Laurent图书馆打造成为一个知识天堂，同时也是一个可以尽情放松身心、放飞梦想的文化圣地。

该图书馆呈分裂状的几何造型便于其和谐地融入周围的森林，特别值得一提的是，这里的三角叶杨有长达20年的生命期。为了进一步促进生物多样化，建筑师在此种植了种类丰富的植被，并且一直延伸至Boulevard主街区，这样，该图书馆就成为整片森林的核心。

森林为Saint-Laurent图书馆提供了绿色能源，并作为天然的空气净化器和噪声过滤器。所有LED灯光均来自于太阳能。建筑师在潮湿的区域中安装了高效的排水系统。站在明亮、洁净的窗前，可以将窗外茂密的森林美景尽收眼底。该图书馆能够令公众在其中读书、放松、娱乐的同时，也时刻践行"可持续发展的绿色理念"，为Marcel-Laurin公园申请"绿色节能型主题公园"助一臂之力。

另外，该图书馆兼具博物馆的功能，其大厅用来陈列城市艺术品，结构简洁，功能多样，自然与历史气息并存。

第二大厅中的期刊和书籍出售空间，气氛更加活跃。这里俨然成为市民挑选图书和文化交流的首选之地，成为该图书馆重要的身份标志之一。

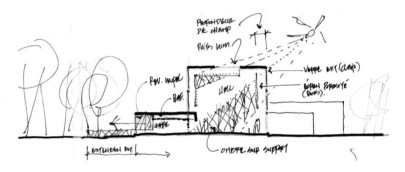

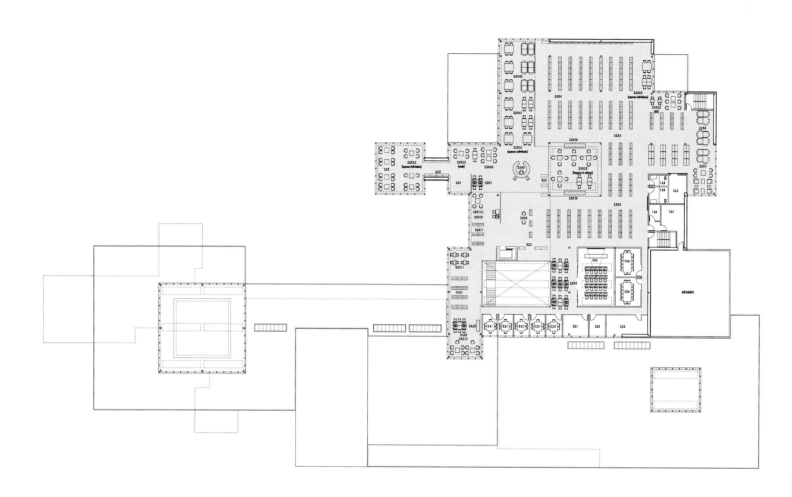

second floor plan
二层平面图

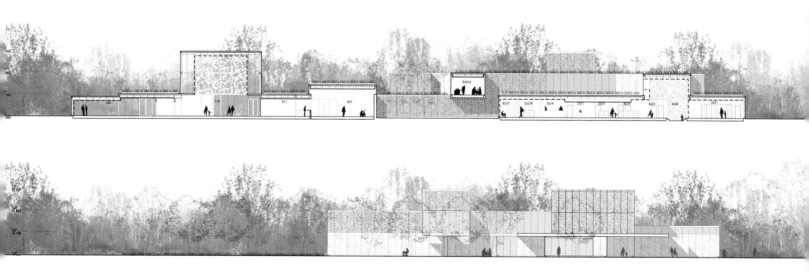

elevation
立面图

Architect: Chevalier Morales Architectes

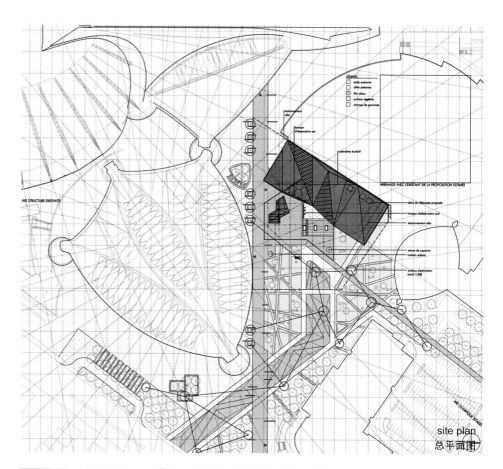

site plan
总平面图

EXPOSING THE INTANGIBLE: GOING BEYOND WHITE BLINDNESS

The project takes its first inspiration in the very particular artificial site in which it is set. Through the complex network of existing structures, the planetarium blends into the white universe of the Olympic installations. In contrast to the opaque and matt concrete used for the Olympic stadium and for the cycling installations, the planetarium is translucent and milky.

With human activity, the interior will be invaded by colors and the many polished and reflective surfaces, like opalescent glass, stainless steel, white perforated aluminium and frit glass, will contribute to amplify this effect. Once inside, articulated volumes and spaces will reveal the object of the museum: the Star theatres, covered in perforated brass panels.

Inspired by today's climate changes and melting arctic ice, the building resembles a tormented and cracked rectangular volume. Slowly sinking next to the Olympic stadium on one side, it seems to be in a precarious balance on the other.

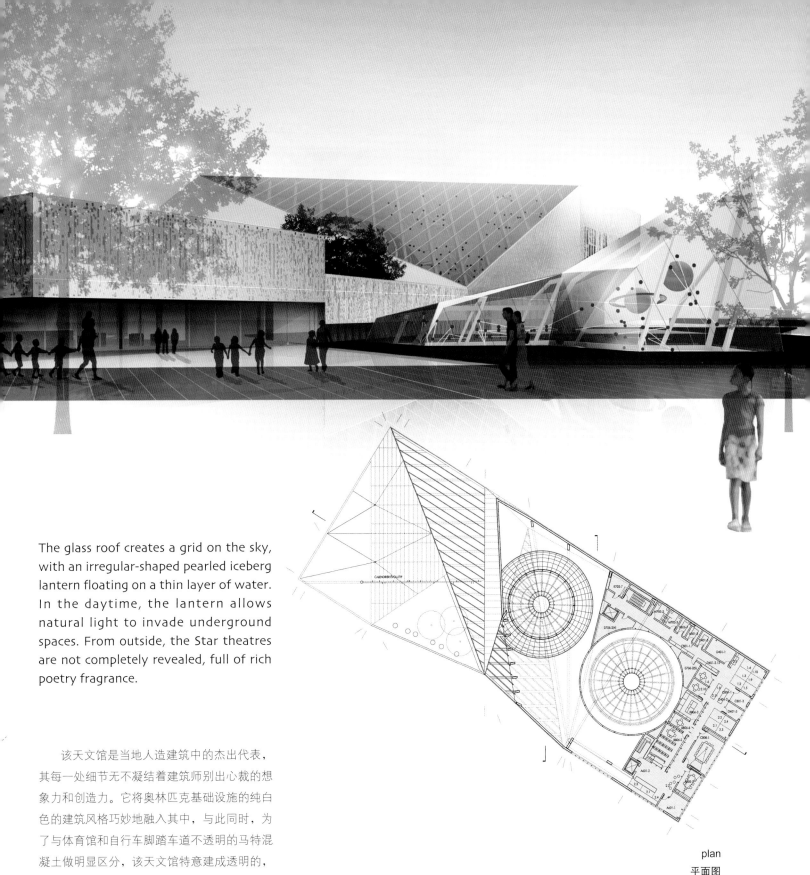

The glass roof creates a grid on the sky, with an irregular-shaped pearled iceberg lantern floating on a thin layer of water. In the daytime, the lantern allows natural light to invade underground spaces. From outside, the Star theatres are not completely revealed, full of rich poetry fragrance.

plan
平面图

该天文馆是当地人造建筑中的杰出代表，其每一处细节无不凝结着建筑师别出心裁的想象力和创造力。它将奥林匹克基础设施的纯白色的建筑风格巧妙地融入其中，与此同时，为了与体育馆和自行车脚踏车道不透明的马特混凝土做明显区分，该天文馆特意建成透明的，并以牛奶色着色。

室内装修创意十足，建筑师运用丰富的色彩和许多反射力很强的、抛光的建筑材料，例如，乳白色玻璃、熔块玻璃、不锈钢、白色穿孔铝合金，这些将大大增强该天文馆透明的"灵性"。一进入室内，首先映入眼帘的便是该天文馆特色十足的"星星剧院"，由穿孔铜板镶嵌而成，光彩照人，夺人眼目。

受到当今气候变化和冰雪融化的启发，该天文馆被建造成一块儿破裂的、不规则的长方形"冰块儿"，恰好和其旁边的体育馆在位置和高度上保持平衡。

玻璃屋顶上有一个高大的露台，上面悬挂着一个仿佛冰块儿一般晶莹剔透的灯笼，与水中的倒影相映成趣。在白天，灯笼将户外的自然光线引入室内；从外面看，灯笼位置居中，使"星星剧院"半开半闭，仿佛少女般"犹抱琵琶半遮面"，充满浓浓的诗意。

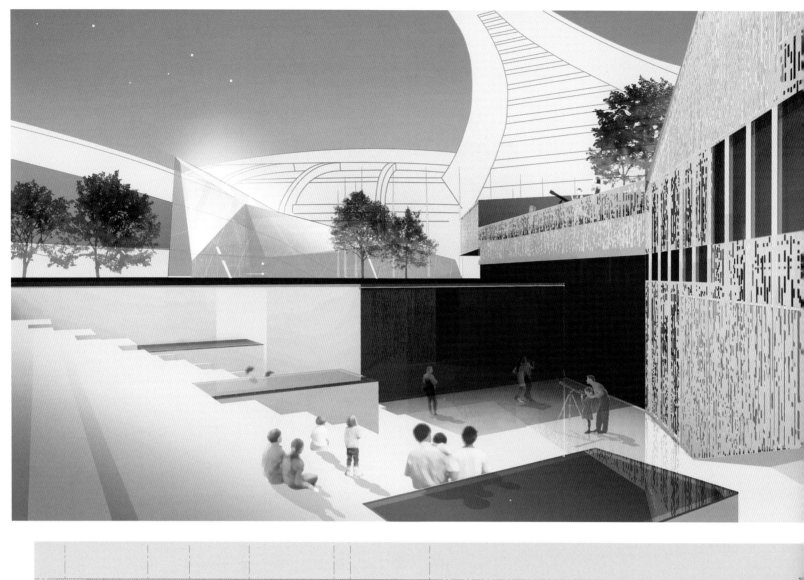

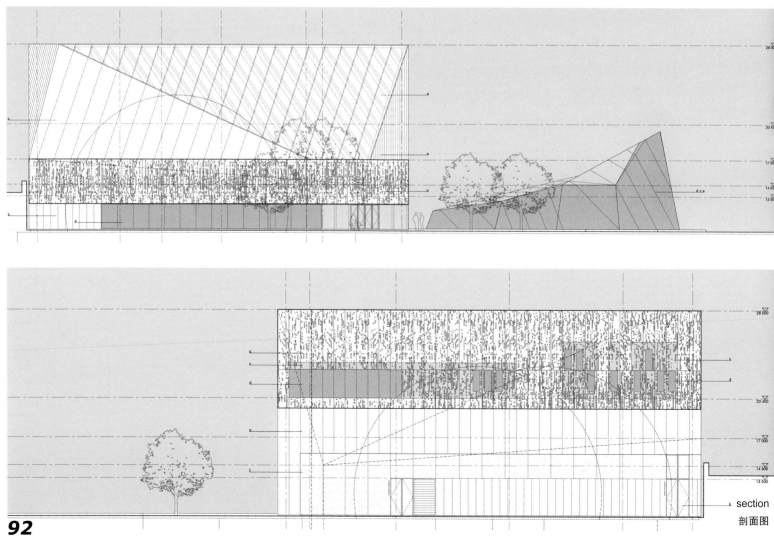

section
剖面图

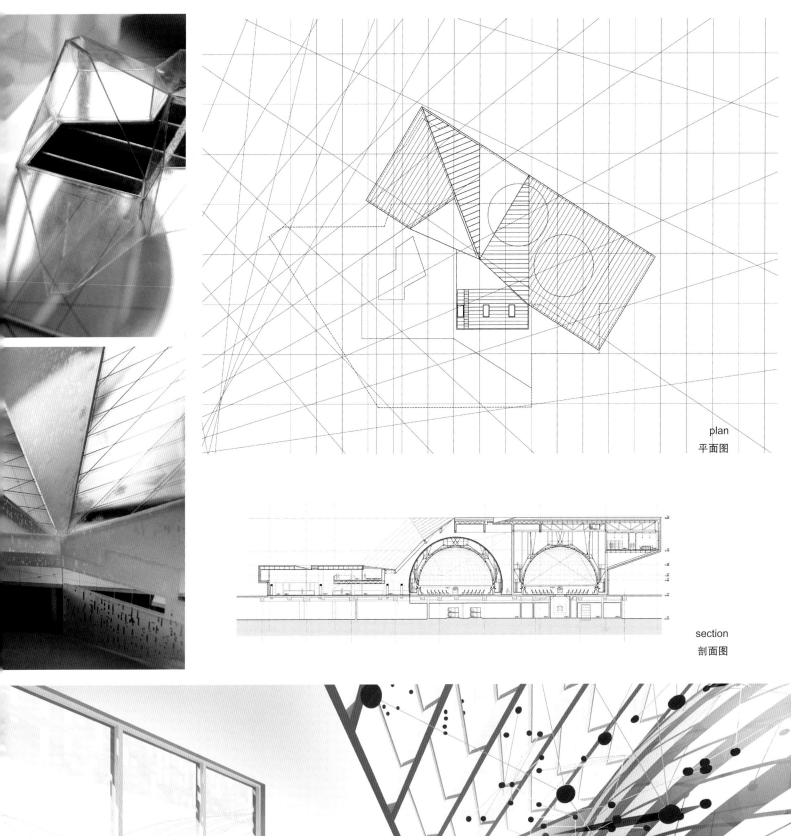

plan
平面图

section
剖面图

Architects: Tétreault Parent Languedoc, Saia Barbarese Topouzanov Architects

UQAM CAMPUS

The original master plan for the UQAM Campus, prepared in 1991, invoked the image of a traditional campus, that of a central green space surrounded by buildings, according to the classic model of several universities. This layout enabled the complex both to fit into its environment and set itself apart from it, while providing the tranquility needed for pursuing studies and research.

UQAM校园的最初设计稿备案于1991年，采用的是传统大学的形象设计，即由建筑物环绕绿地而建，与其他几所大学的经典模式相一致。该布局使建筑群既与周边环境融为一体，又与环境相区别，从而为师生的刻苦学习和专心研究打造一个安静的环境。

The original idea of a green space was transformed into a continuum of yards and gardens enclosed inside buildings or interspersed between them. The green space also includes paths and a pedestrian walkway set up along the forgotten traces of what was once Kimberly Street and the extension of Evans Street. Nature has insinuated herself into these man-made structures. All these spaces come together, giving structure to the architecture and bathing

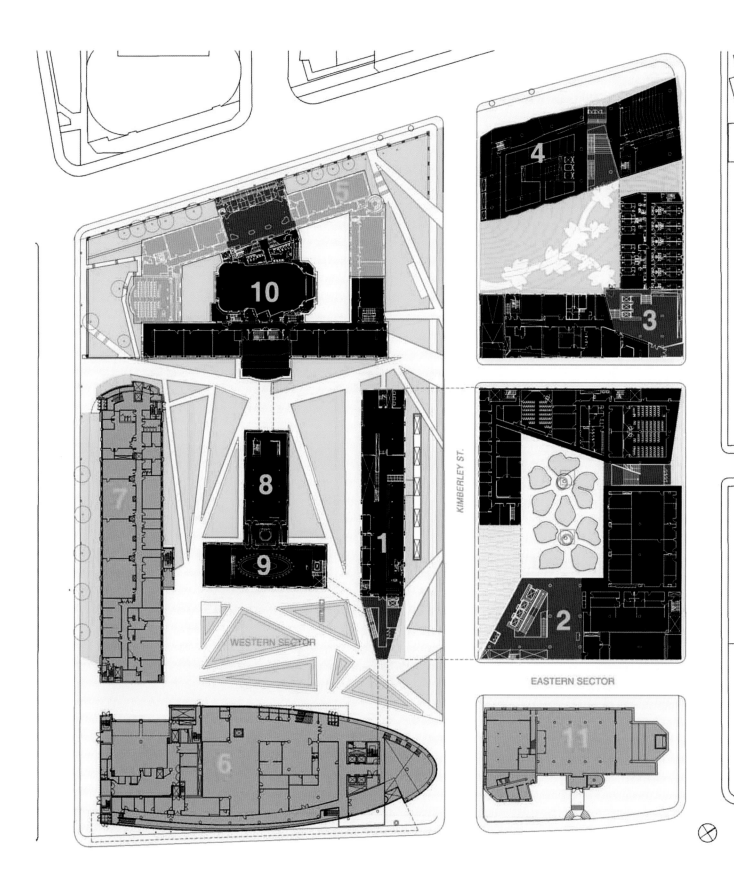

1. central library
2. biological science building
3. student residence
4. broadcasting building
5. sherbrooke building
6. president-kennedy building
7. biochemistry building
8. mediatech
9. student cafe
10. new auditorium
11. st-john the evangelist church

■ existing building
■ intervention
■ interior public spaces

plan
平面图

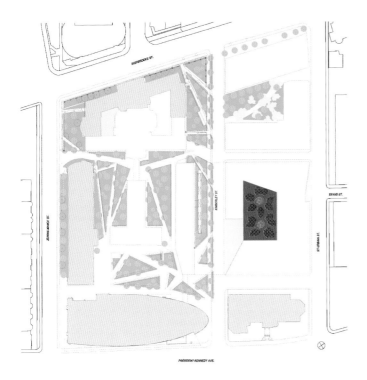

landscape plan
景观平面图

them in light. Their intertwined networks link street accesses to the entrances of the various pavilions. The spaces provide intimate settings, which are ideal for meetings, discussions, relaxation and reflection. The busy paths are lined with rows of trees, whose irregular patterns evoke the forest. In the gardens, giant flowers, some of them made of slate, pave the walkways between the buildings (student residences' garden), while other form petals around a tree assume the role of a giant pistil (Biological Sciences Pavilion yard). Indigenous species or plants of botanical interest have been selected for their ability to adapt to the urban climate. They introduce greenery within the city, in a neighborhood that is in desperate need of foliage.

In short, the traditional shape of campuses is evolving: the new configuration is one in which spaces overlap, and where spaces and erected structures are architectural equals.

最初设计中的绿色空间转变成了整体、连续的庭院园林，或被楼群环绕，或镶嵌其中。其绿色空间包括幽幽小径、建在被遗忘的Kimberly大街遗址上的人行道，以及Evans大街的延伸部分。大自然巧妙地潜入人造建筑的缝隙。所有空间组合在一起，构成该建筑的全体结构并形成一个个采光区。网状的街道连接着各处楼宇。该空间中还留有私密的场所，供人在此进行会议、研讨、放松和思考。熙熙攘攘的小径旁，一排排树木以其不规则的图案激发了人们对森林的渴望。在花园中，硕大的花朵图案由石板铺设而成，在两楼之间架设了几条通往学生寝室的小路，其他形状的翼瓣环绕着树木，仿佛巨大的花蕊一般。这里还种植着一些本地的树种或具有研究价值的植物，因为它们比较适合该城市的气候。它们将草木的绿意引入城市中，美化了渴望变得枝繁叶茂的周边区域。

总之，UQAM校园对传统型的设计进行了改良：将空间重叠，使平面结构与垂直结构相结合，以达成建筑造型的平衡。

Architect: WZMH Architects

Landscape Architect: Quinn Design Associates Interior Designer: Cannon Design

Structural Engineer: Halsall Associates Limited

Location: Oshawa, Ontario, Canada

Area: 41,957 m²

Photographer: Shai Gil

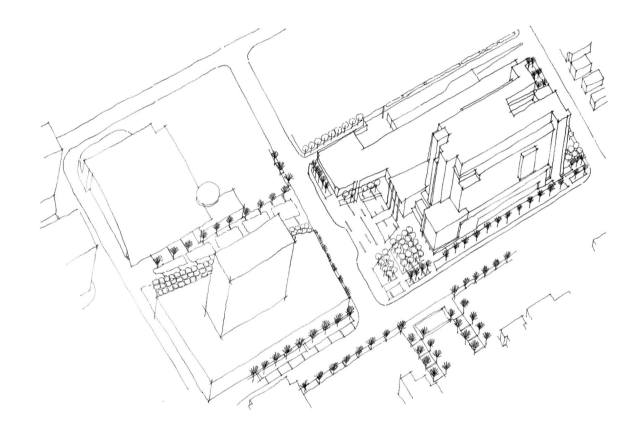

DURHAM CONSOLIDATED COURTHOUSE

With its richly patterned cladding of spandrel, clear glass massing, Durham Consolidated Courthouse, completed in January 2010, makes a significant contribution to the emerging urban framework of downtown Oshawa. Its bold, modern vocabulary emphasizes transparency and openness both for users and passersby.

A large outdoor public space, Courthouse square, is the forecourt to the building entrance. The scale of the main entrance pavilion on the square establishes a sense of dignity, appropriate for the front door of a courthouse.

Providing badly-needed space for the province´s judicial system, this six-storey, 40,000 m² structure houses 33 courtrooms, associated support space and prisoner-holding facilities.

完成于2010年的Durham Consolidated法院，运用一系列种类丰富的、透明的拱肩式玻璃，这对于丰富都市建筑框架具有至关重要的作用。该法院大胆、新潮、现代化的建筑语汇，向公众充分展示了"透明和开放"。

走进该法院，首先映入眼帘的是一片巨大且开阔的空地。其中央矗立着一个笔直、挺拔的凉亭，庄重、威严，仿佛一名高大威猛的战士，在勇敢地捍卫着自己的尊严。该法院分为6层，占地4万平方米，拥有33个房间，分别用于举行各类审判、提供司法诉讼服务，以及存放监狱设备。

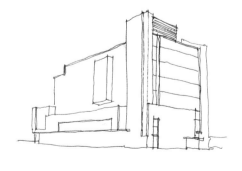

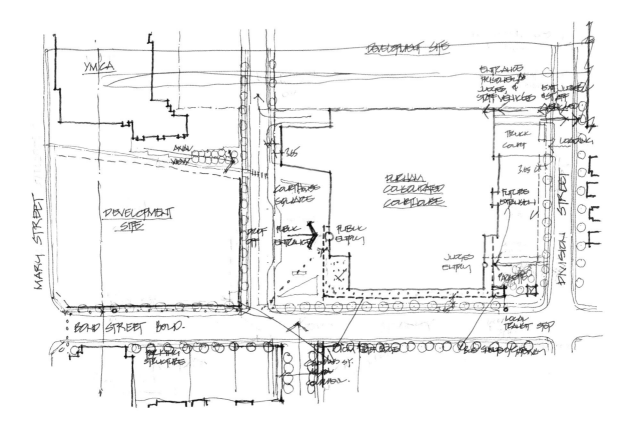

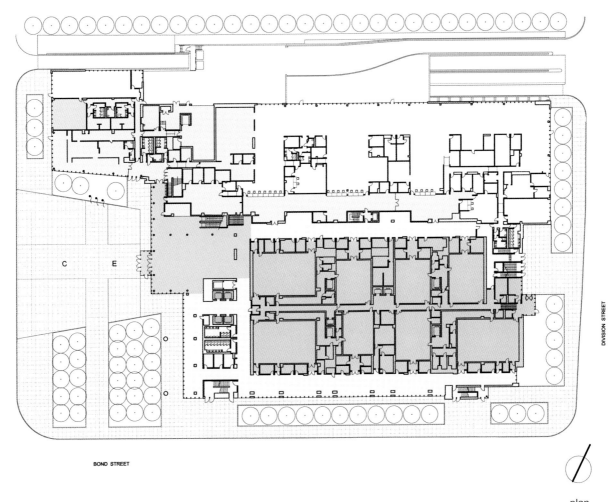

plan
平面图

109

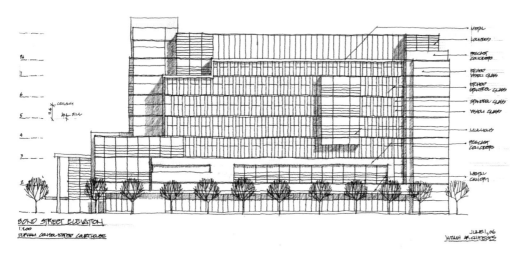
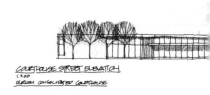

elevation
立面图

The courthouse represents a physical embodiment of our justice system interpreted in a modernist language. Solid building elements serve to express the stability and permanence of the courts. Welcoming to the public, the courthouse elevations are highly transparent, using clear glass with a rich mosaic of white ceramic frit glass panels. The scale of the main entrance pavilion creates a formality and sense of dignity, appropriate for the front door of a courthouse.

On the west, the main building core is raised to create a visual terminus, looking east from the proposed linear park. At the southeast corner of the site, a column of glass that is illuminated at night creates a strong gateway to downtown Oshawa.

The project exhibits a new typology for a typical courtroom floor that has a "back to back" arrangement of courtrooms that results in short walking distances for judges and staff. The availability of daylight and views to the outside in the courtroom waiting areas will reduce stress for the participants in court proceedings.

该法院坚固的建筑外观代表了司法的稳定性和永久性。电梯由清澈的嵌花式白色陶瓷玻璃制成，乘客在其缓缓上升的过程中能够看到各层的办公情景，影射了透明和开放的司法原则。

西侧的建筑突然被抬高，站在此处，向东可以鸟瞰即将建成的线性公园；在西南一隅，一排排明亮的玻璃在月夜中闪闪发光，照亮了通往市中心的道路。

该设计提供了一个法院建筑的新式模型："背对背"的房间缩短了各个部门之间的过道距离，方便工作人员的沟通与协作；良好的采光和美丽的风景大大缓解了法律诉讼程序过程中的紧张气氛。

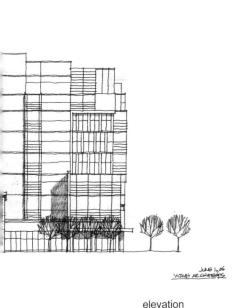

elevation
立面图

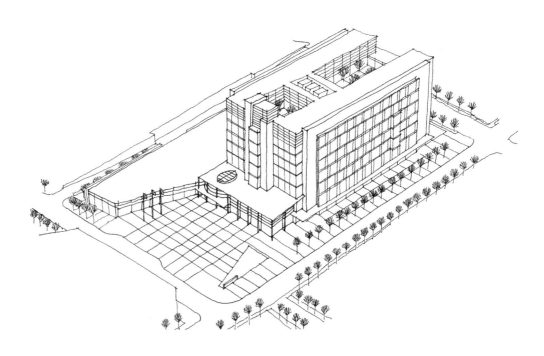

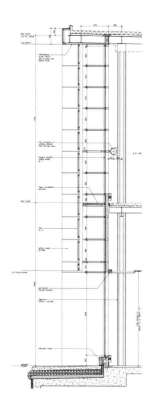
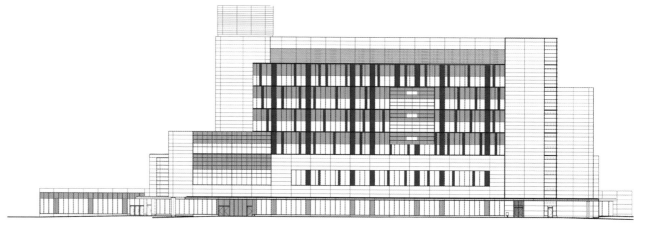

south elevation
南立面图

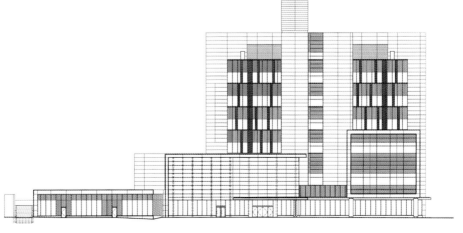

west elevation
西立面图

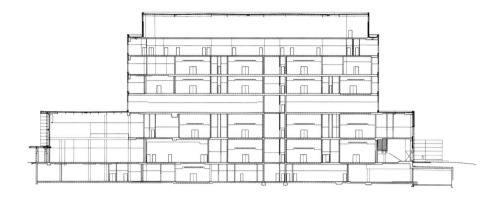

east-west section
东西剖面图

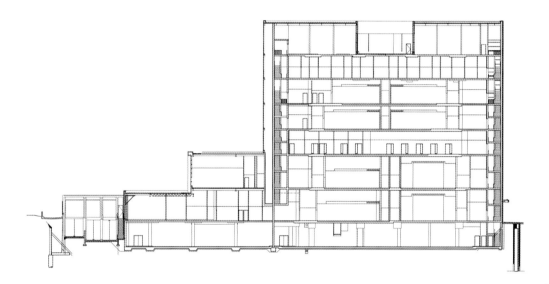

north-south section
南北剖面图

Architect: Kleinfeldt Mychajlowycz Architects Inc.
Location: Brampton, Ontario, Canda
Area: 20,438 m²
Photographer: A-Frame Studio

ROY MCMURTRY YOUTH CENTRE

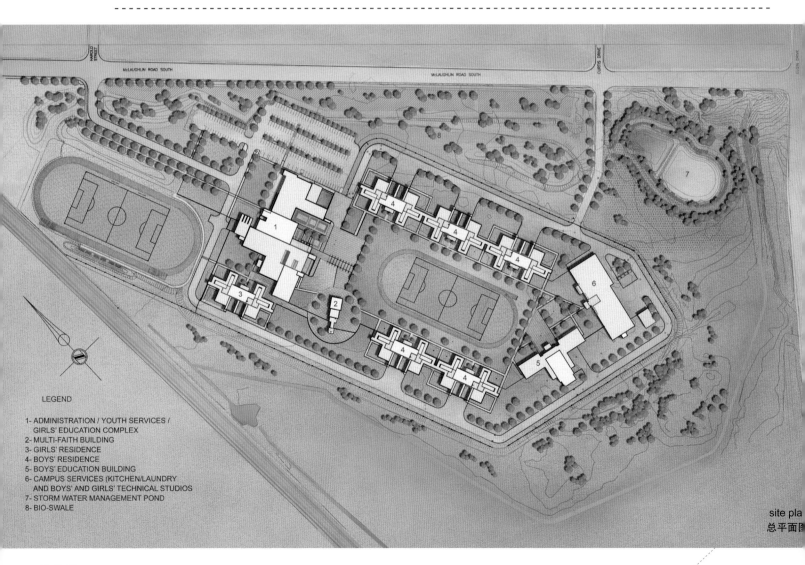

LEGEND

1- ADMINISTRATION / YOUTH SERVICES / GIRLS' EDUCATION COMPLEX
2- MULTI-FAITH BUILDING
3- GIRLS' RESIDENCE
4- BOYS' RESIDENCE
5- BOYS' EDUCATION BUILDING
6- CAMPUS SERVICES (KITCHEN/LAUNDRY AND BOYS' AND GIRLS' TECHNICAL STUDIOS
7- STORM WATER MANAGEMENT POND
8- BIO-SWALE

site plan
总平面图

The decision made in 2004 by the Provincial Government of Ontario to introduce a new Ministry of Child and Youth Services, as distinct from the adult facilities, has allowed a re-evaluation of the programs and physical expression of them in consideration of youth offenders.

The site is an existing 400,000 m2 institutional site which was initially surrounded by agricultural fields. The site is cleared of an existing women's prison, with the exception of two buildings, which were maintained and renovated as part of the Youth Centre. Eight new buildings have been added to the site to provide services to 192 youth offenders, 32 girls and boys, aged 12 years to 17 years old. Over 300 staff members, volunteers and family members and visitors are included in the immediate community of the centre.

The secured area of the site is a Campus Morphology, reinforcing the education ideals of the centre. A garden is located in the centre. Contiguous buildings, infill board-formed concrete, masonry panels and Corten steel panels define the garden wall. The Campus is surrounded by a public park, offering a naturalized landscape including a bio-swale, pond, existing mature trees, hundreds of new trees and a new streetscape. Campus Morphology is critical to the success of this Youth Centre, both in promoting a healthy public understanding of the ideals of the institution and the residents understanding of their place and obligations in the society.

2004年，安大略省政府决定成立"青少年服务部"。这一举措也促成了Roy McMurtry监狱的整修——新建一个青年活动中心，拓宽原有场地，并对现有项目进行重新规划，这一切旨在为这里的青少年提供不同于成年人的、更有针对性的服务。

40万平方米的McMurtry监狱周围分布着大量的农业用地。这次整修，将原先女子监狱的两栋建筑拆除，在此基础上，新建了8栋大楼，作为青少年活动中心，为这里的192名青少年提供服务，其中包括32名12至17岁的孩子。该活动中心能够容纳超过300人，其宽敞的空间旨在邀请这

里的工作人员、志愿者以及青少年的亲属参加丰富多彩的、即兴的社区活动。

该活动中心的一大亮点是运用"校园形态学"的建筑理念，强化该活动中心的教育功能。庭院中的建筑物呈连续性分布，庭院中央有一个花园，填充木板式混凝土、石工技艺镶板，以及柯尔顿钢板，共同组成了花园的墙面。庭院被其外部的公园所包围，拥有两个沼泽、一个池塘、一排排成熟的老树和刚刚种植的新树，形成一道亮丽街区风景线。"校园形态学"的完美展现，旨在使这里的青少年对"制度化建设"有一个正确的理解，并逐渐意识到他们所扮演的社会角色和应当履行的社会职责。

Architect: Consortium KPMB FSA
Location: Montreal, Quebec, Canada

JOHN MOLSON SCHOOL OF BUSINESS

After opening its doors officially for the first time on September 22, John Molson Business School unveiled its new facilities to students and public alike in the "Quartier Concordia" area of Downtown Montreal. The firms of Kuwabara Payne McKenna Blumberg Architects and Fichten Soiferman Associates formed a consortium, KPMB / FSA, for this leading green project. This project boasting world class environmental technology, as well as state-of-the-art facilities for faculty and students is part of a three-phase project including the Guy Metro Building and the Engineering-Computer Science and Visual Arts Integrated Complex. All the three are inter-linked underground and connected to the STM Metro system.

Within its two sub-basement levels and 15 levels above ground, there are 45 state-of-the-art classrooms, a 300-seat auditorium, two 150-seat amphitheaters, four 120-seat amphitheaters, 22 conference rooms, and numerous other office spaces and services housed within the complex.

This exemplary building includes many features that have contributed to its candidacy for a LEED designation, which is currently under review. Many components ranging from the use of durable materials like granite, metal, glass and ceramic, and the use of high performance thermal glass on its exterior address this "green" sensibility in very practical ways. In addition, KPMB / FSA have integrated a green roof on the 4th floor terrace. What's more, the architects have also incorporated solar panels that were developed at Concordia, on the top of one of the facades of the large structure. This solar wall produces enough energy to significantly reduce the energy consumption of the building; and is equivalent to heating seven Canadian homes throughout the year.

John Molson商学院于9月22日正式成立，向蒙特利尔市中心Quartier Concordia地区的市民解开了其神秘的面纱。Kuwabara Payne McKenna Blumberg 设计院和Fichten Soiferman协会组成了一个强有力的财团——KPMB / FSA，领导该项绿色工程的实施。该建筑以其世界顶级的环保技术和先进的设备而著称，分为三个组成部分：Guy Metro大楼、工程计算机科学教学楼和视觉艺术大楼。它们彼此相连，共同组成了STM Metro系统。

两层的地下空间和15层的地上空间，分布着45间拥有先进设备的教室、一个拥有300个座位的礼堂、3个露天剧院（两个拥有150个座位，一个拥有120个座位）、22间会议室，以及宽敞、明亮的办公空间。

该建筑已经申请LEED奖项的评选，并且胜券在握。使用永久性建筑材料，例如，花岗岩、金属、玻璃和陶瓷，使用高性能的热玻璃建造其外观，这些都充分表明其所独有的"绿色可操作性"。另外，建筑师在其天台上修建了一个绿色的屋顶。特别值得一提的是，建筑师在其中一个外立面上安装了本地生产的太阳能镶板，它在提供足够能源的同时，也降低了该建筑的能源消耗。该太阳能镶板所产生的热量，足以为7个加拿大家庭提供全年的供暖。

Architects: Diamond+Schmitt Architects Inc., Provencher Roy+Associates Architects

MCGILL UNIVERSITY LIFE SCIENCES COMPLEX

The primary function of the McGill University Life Sciences Complex is research in cancer and biomedicine. This includes five key components: chemical biology, complex traits, developmental biology, cell information systems and cancer research. The Complex integrates the new facilities, the Francesco Bellini Life Sciences Building and the Cancer Research Building, as well as the existing McIntyre Medical Sciences and Stewart Biological Sciences buildings.

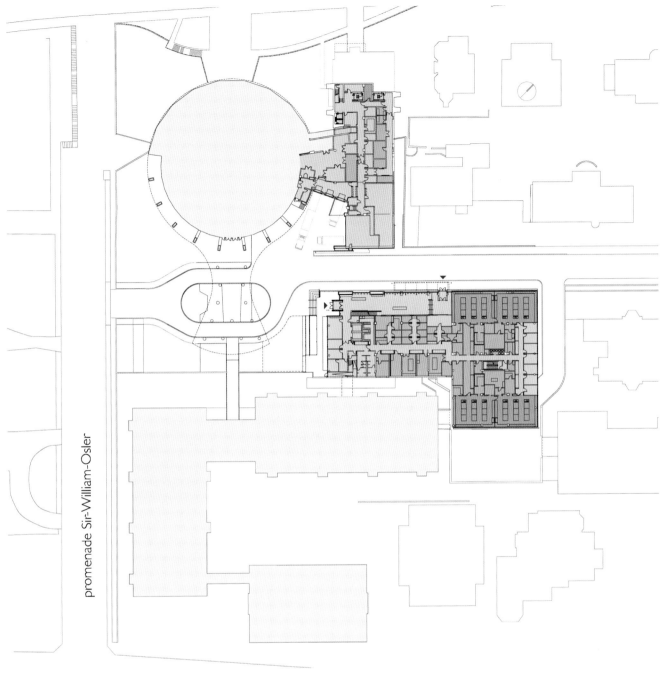

promenade Sir-William-Osler

Linking elements and informal social spaces tie the new facility to both adjacent buildings. At the upper levels, this allows for research inter-connectivity. At the lower, more public levels, these spaces encourage casual interaction between users during breaks. The four-storey interior atrium space doubles as a pedestrian passage, leading to vertical circulation into the complex and enhancing social and academic campus life.

McGill大学生命科学院的主要研究领域为癌症学和生物医药学，具体而言，包括五个方面：化学生物学、复合特征、发展生物学、细胞信息系统以及癌症研究。这次整修，在原有生命科学院主楼（McIntyre医药科学院和Stewart生物科学院）的基础上，新建了Francesco Bellini生命科学大楼和癌症研究中心。

通过线条简洁、宽敞、明亮的人行通道，建筑师巧妙地将各个建筑之间的空间连接起来。这样，在高层，方便了各研究部门之间的交流与协作；在低层，使员工能够在休息时更加随意、放松。4层高的内部中庭，扩大了人行通道，一直通向生命科学院主楼的垂直流通空间，提升了校园社会生活和学术生活的质量。

129

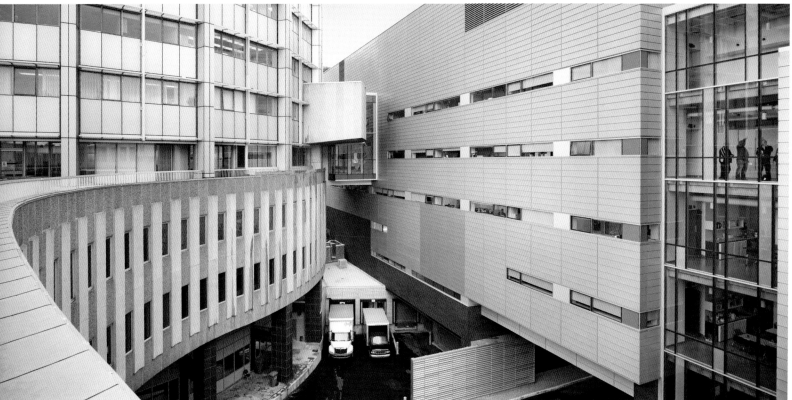

The Life Sciences Complex is sited adjacent to one of the most cherished green spaces in Montreal – the upper slopes of Mount Royal. The sensitive context, coupled with the University's sustainable building mandate and the architects' commitment to reducing the ecological impact of architecture, helped to establish the design team's goal of constructing an unobtrusive energy efficient building. The new Bellini and Cancer pavilions are designed to achieve LEED Gold certification with the Canadian Green Building Council.

An integrated design approach, including value management sessions, have been utilized to ensure architectural integration of sustainable design features and to fully understand the impact each decision would have on operating and maintenance costs during the life of the building. Each energy conservation measure was considered individually, based on a 10-year payback benchmark. Overall, the total building will use 53kWh/m2 annually, 36% more efficient than the Canadian National Model Energy Code reference building, deserving to be the exemplary model of the energy-efficient architecture.

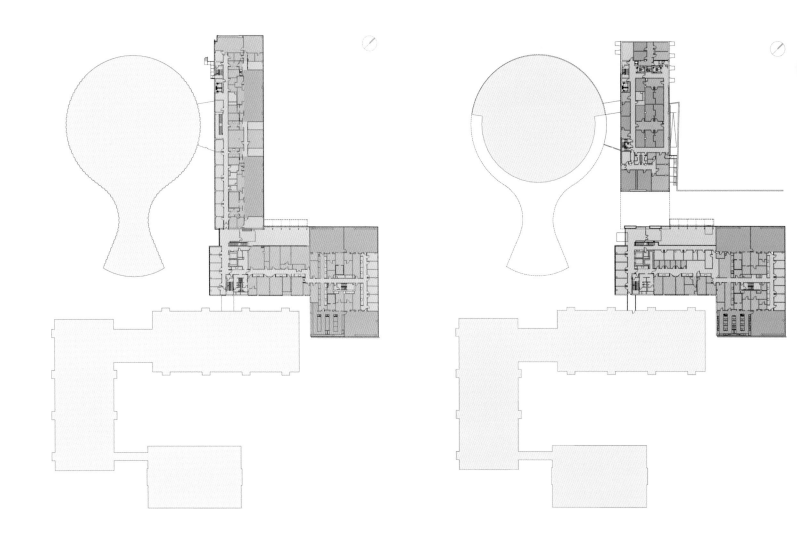
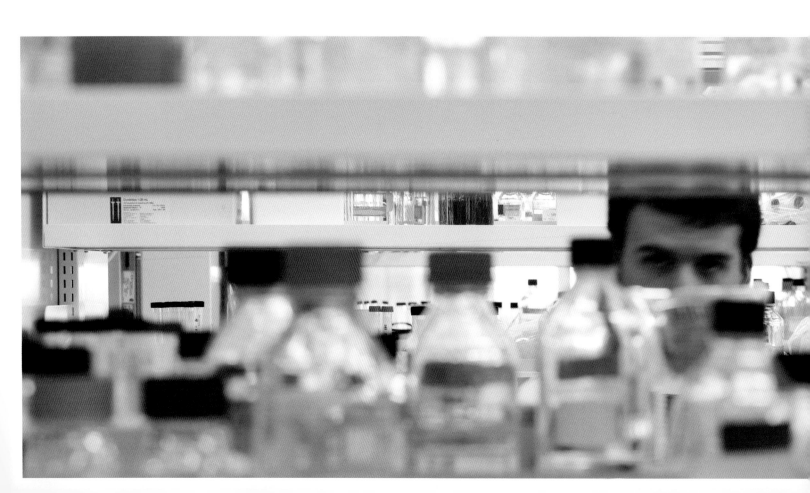

　　McGill大学生命科学院毗邻蒙特利尔最著名的绿地——Royal山脉的上层斜坡。敏感的周边地理环境、McGill大学建造可持续性建筑的任务，以及建筑师降低该建筑的环境影响的决心，这一切都促成这些"能源高效型"建筑的诞生。新建的Francesco Bellini生命科学大楼和癌症研究中心则有望获得由加拿大绿色建筑委员会颁发的LEED金奖。

　　该设计过程始终坚持"价值管理评估"，对每一栋建筑在未来10年内的"成本价值损耗"进行严格和精准的测算，以确保每一栋建筑具有可持续性，并对其需要的运营和维修费用，以及其对周边环境造成的积极或消极的影响，进行综合性评估，最终得出具有统计意义的数据。总体而言，McGill大学生命科学院的建筑每平方米每年将耗能530千瓦·时，相比加拿大国家建筑能源消耗标准，节约了36%的能源，堪称"能源高效型"建筑的典范。

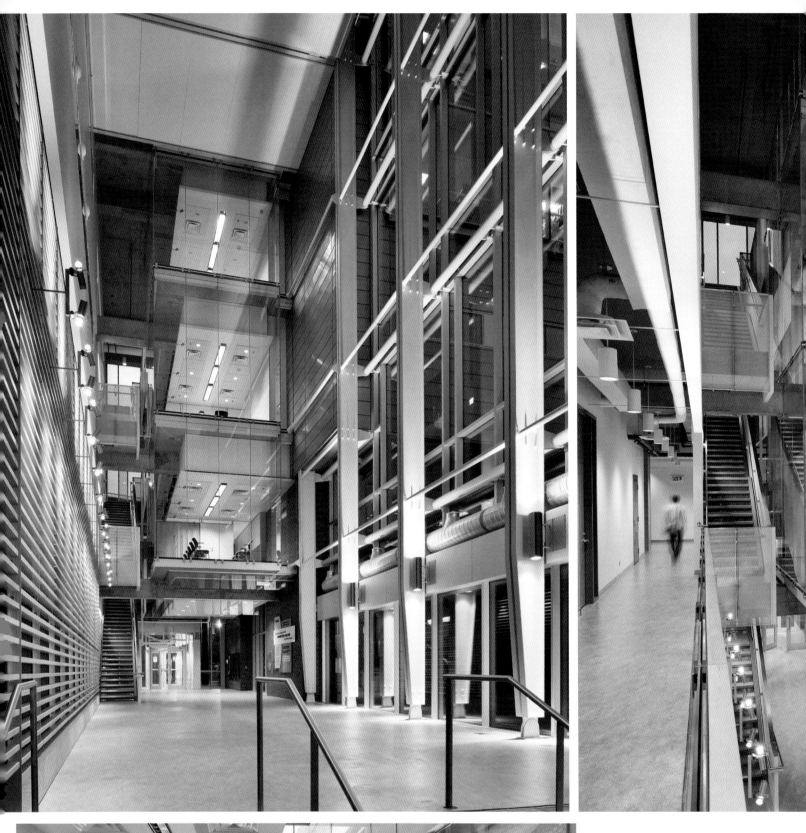

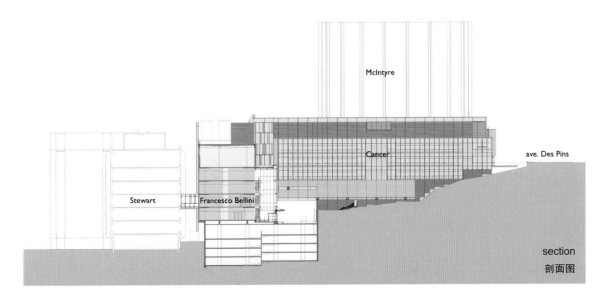

section
剖面图

Architects: Fournier, Gersovitz, Moss, Drolet et Associates Architects

D'ÉTUDES NORDIQUES COMMUNITY SCIENCE CENTRE

The centre is devoted to scientific research and exchange of traditional knowledge.

One of the most significant part of the centre is that a hall was designed specifically to host school groups of all levels for teaching activities and popularization of science; the hall will also display a permanent exhibition describing the territory's cultural and scientific history.

D'Études Nordiques社区科学中心致力于从事科学研究和传统知识的传播。

该建筑最主要的构成要件是一个大厅，用来组织各种各样的教学活动和学术展览。可以说，这里就是该地区文化和科学活动的中心区域。

该建筑最大的亮点在于严格遵循了可持续发展原则和环境保护原则。它地处加拿大北部地区，这里气候寒冷，自然条件相对恶劣，生态环境极其脆弱，交通运输受到气候条件的极大限制，工程施工时间相对较短。然而，FGMDA建筑设计事务所的建筑师就地取材，充分利用太阳能供暖。南北朝向的窗户、提供地热的木质地板，以及光电结合的建筑设备，每一处细节都是建筑师的巧妙构思和精心设计。

该建筑的外观材质高度保暖，当气温下降至-40°C以下时，"密封功能"自动开启。建筑框架由木材构成，例如，黑色分层云杉木、红色雪松木、黄色桦木等。另外，每一个入口都设立了两个前厅，这样能够有效地阻挡户外凛冽的寒风，有利于保暖。

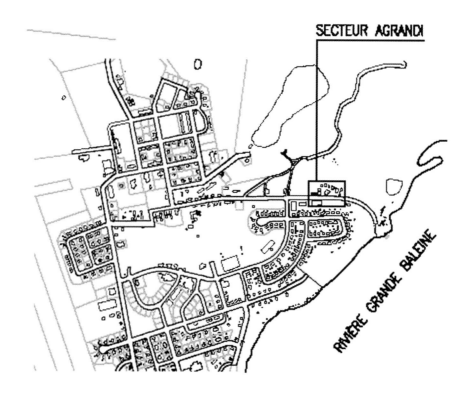

Sustainable development and conservation of nature are not new practices, especially in a hostile northern environment devoid of resources and fragile. Transportation of materials is difficult, the construction season is short, and climatic constraints impose specific logistical solutions. Well versed in construction in northern regions, FGMDA designed an efficient building that respects the environment and reduces energy consumption thanks to passive solar heating, abundant south-facing fenestration, floors used as thermal mass, and integration of photovoltaic cells.

An exterior envelope presenting high levels of thermal resistance and airtightness is essential when temperatures drop below -40°C. Wood is used for both the frame and the exterior envelope (red cedar cladding, pine substructure, and yellow birch panelling). In addition, the integration of a double vestibule at each entrance reduces air exchanges and helps to conserve energy.

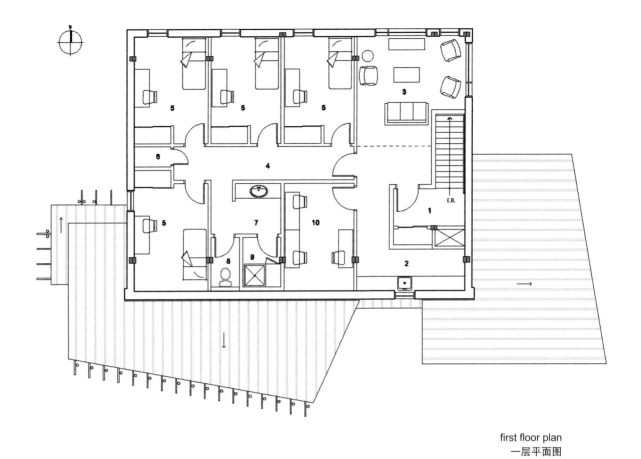

first floor plan
一层平面图

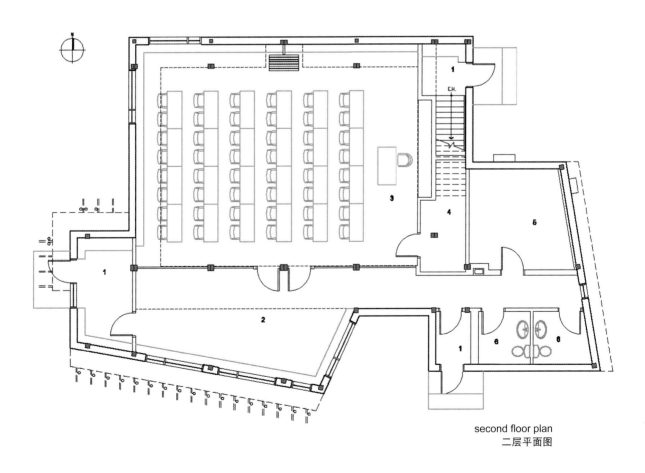

second floor plan
二层平面图

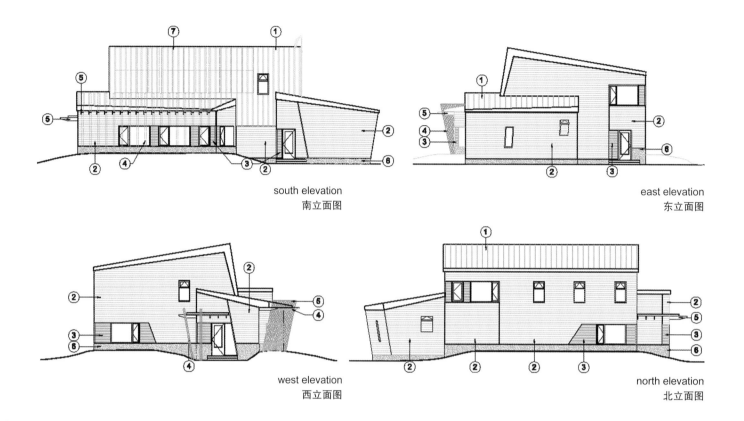

south elevation
南立面图

east elevation
东立面图

west elevation
西立面图

north elevation
北立面图

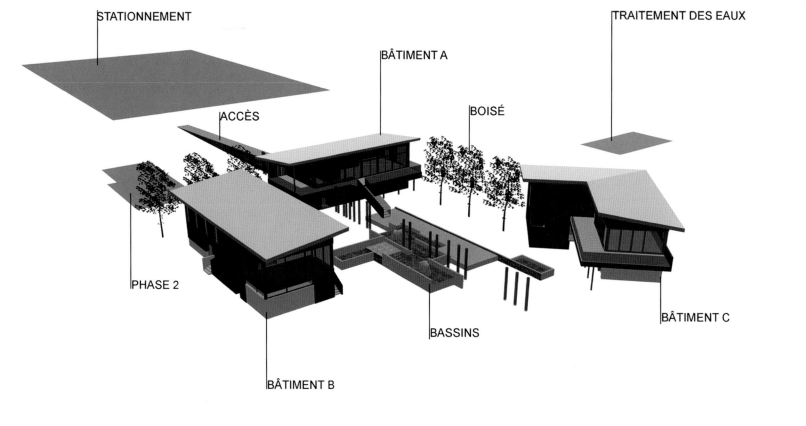

146 LUXURY, SERENITY AND INNOVATION: LE GERMAIN CALGARY HOTEL
152 STRÖM SPA NORDIQUE
164 BRANDON NO.1 FIREHALL
170 LES LOFTS REDPATH
178 MANITOBA HYDRO PLACE
184 CIDRERIE LA FACE CACHÉE DE LA POMME
190 REFURBISHMENT AND EXTENSION OF THE CRÉVHSL HEADQUARTERS
200 STATION BLÜ

Architect: Lemaymichaud Architecture Design

LUXURY, SERENITY AND INNOVATION: LE GERMAIN CALGARY HOTEL

The Le Germain Calgary Hotel boasts an exceptional and excellent location in the heart of downtown. Connected to the city's network of elevated pedestrian walkways, the hotel, unlike other Germain hotels, actually opens directly on to 9th Avenue. The lobby features a two-story glass wall facing the main commercial street, thus creating a unique vibrancy between the street and the hotel space.

Le Germain Calgary旅馆坐落于城市中心的黄金地带，地理位置优越，是众多人行横道的汇集地，且直接通往繁华的第九大道。其内部两层高的玻璃墙，恰好面对着主要商业区，驻足旅馆一隅，就能将玻璃墙外熙熙攘攘的商业世界尽收眼底，美妙无比。

The materials and interior design were selected to combine ease and comfort. "Our goal was to create a welcoming environment with a flash of innovation," the designer explains. The hotel's exterior cladding of Prodema wood-laminate panels prefigures the warmth of the interiors. Beyond the impressive fenestration of the façade, the main entrance is a vibrant, active space, perfectly bringing the beautiful outside world into the hotel.

Between the lobby and front desk area, a wall of recycled felt, in shades of grey and black resembling shale, adds both texture and acoustic properties. The whole hotel as if is sleeping in a cradle, sweetly and happily.

In addition, 90 wells are drilled to supply the hotel with enough geothermal energy to heat the water and some radiant floors. All rooms are equipped with the technological necessities of the modern travelers (e.g., Internet access, flat-screen television), facilitating guests to keep in touch with the outside world.

The hotel embodies a splendid experience of luxury, serenity and innovation.

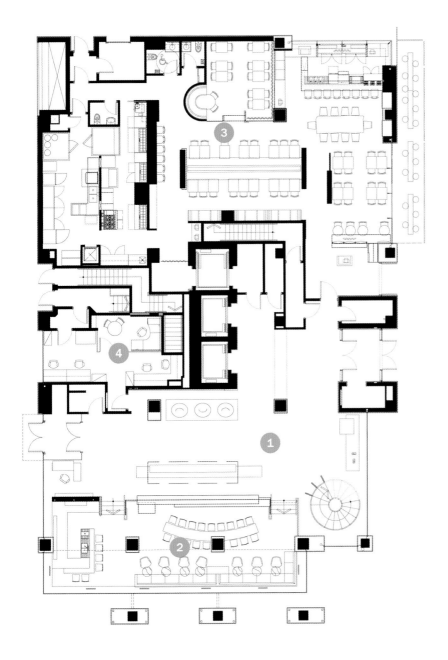

1. lobby
2. lounge
3. restaurant
4. administration

site plan
总平面图

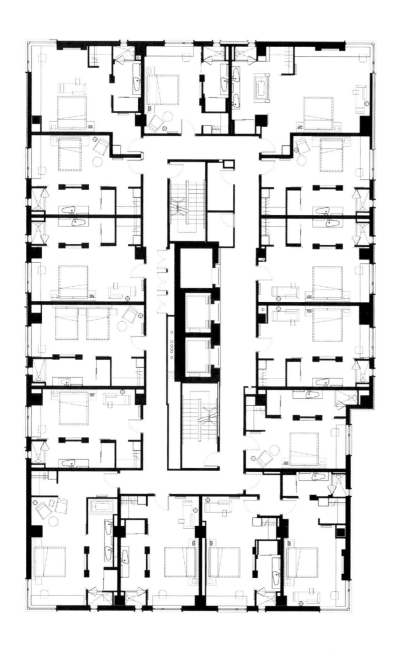

该旅馆的建筑材质和内部设计旨在营造舒适和安逸之感，打造温馨、宜人的居住环境。外部材质则选用Prodema木质花边的钢板，散发着一股令人倍感亲切的自然气息。入口处是一个生机勃勃的空间，装饰奢华的接待大堂，仿佛将室外的自然美景带入室内，并延伸至旅馆中的每一个角落。

大堂以及每个房间之间都设置了由灰色和黑色页岩制成的隔音墙，消声效果极佳，整个旅馆仿佛在缓缓荡漾的摇篮中，静静地进入梦乡。

除此之外，该旅馆拥有自成体系的90口井，为其提供地热能和热水。所有房间均配备了高速上网设施和液晶彩色电视等等，其运用当代先进技术，令旅客在这里能够与世界各地保持联系。

该旅馆为旅客成就了一场奢华、宁静、创新的现代之旅。

floor plan
楼层平面图

Architect: Chevalier Morales Architects
Engineers: Les Consultants Gemec, Génivar
Project Managers: Sergio Morales, Stephan Chevalier
Project Team: Sergio Morales, Stephan Chevalier, Karine Dieujuste, Christine Giguère, Samantha Hayes, Neil Melendez
Location: Montreal, Quebec, Canada
Area: 1,000 m²
Photographer: Marc Cramer

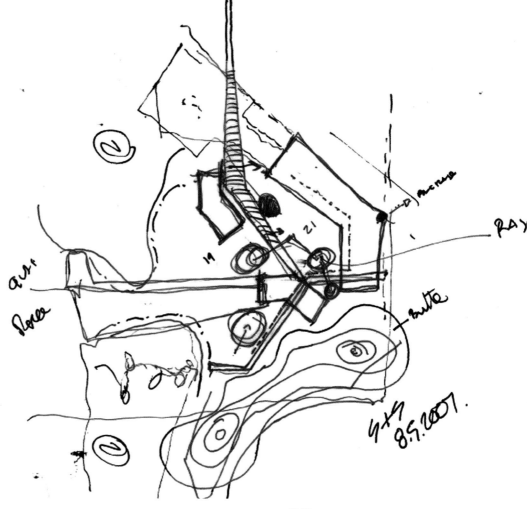

STRÖM SPA NORDIQUE

Exterior spas seem to be taking a more important place in people's life compared to traditional theme parks.

Ström Spa Nordique provides this new type of playground for the body and mind offers a sensitive contemporary design. Three separate buildings are placed delicately on the manmade topography and look out onto the Lac des Battures. The buildings are occupied by saunas, massage and relaxation rooms, Turkish baths and other spaces necessary to make a functional spa. Covered in clay grey brick and white wood, the buildings exude calm and serenity. Brass elements like strange sculptures, gargoyles and interior furniture are making reference to all piping, hid underground or in mechanical rooms, responsible for the magical experience that offers the spa.

At night, the blend of neon lights, LED lighting and flames of torches illuminates the surreal décor, lightly and brightly.

温泉浴场已经深刻地融入现代人的生活,其重要性已经远远超过了传统式的主题公园。

该温泉浴场为人们放松身心提供了一个绝佳胜地。它分为三栋建筑,均建在一座海拔并不高的人造假山上,站在山顶,呼吸着新鲜空气的同时,放眼望去,就能够将Lac des Battures云山雾罩的美景尽收眼底。三栋建筑中分别是桑拿房、按摩室和休息室、土耳其澡堂等等。整个温泉浴场使用灰色黏土瓷砖和白色木头铺设而成,营造了宁静、理性的空间氛围。深处其中,深吸一口气,顿时令人心旷神怡、豁然开朗。室内的布局设计更加别出心裁:造型荒诞的人体雕塑,形状诡异的滴水嘴、石像鬼,还有那些奇形怪状的家具,令人好像在享受着一场神奇的探险之旅。

在夜晚,处于各个角落的霓虹灯、LED灯以及燃烧的火炬,一闪一闪,仿佛一双双晶莹剔透的大眼睛,一眨一眨,点亮了整个温泉浴场,温馨且明亮。

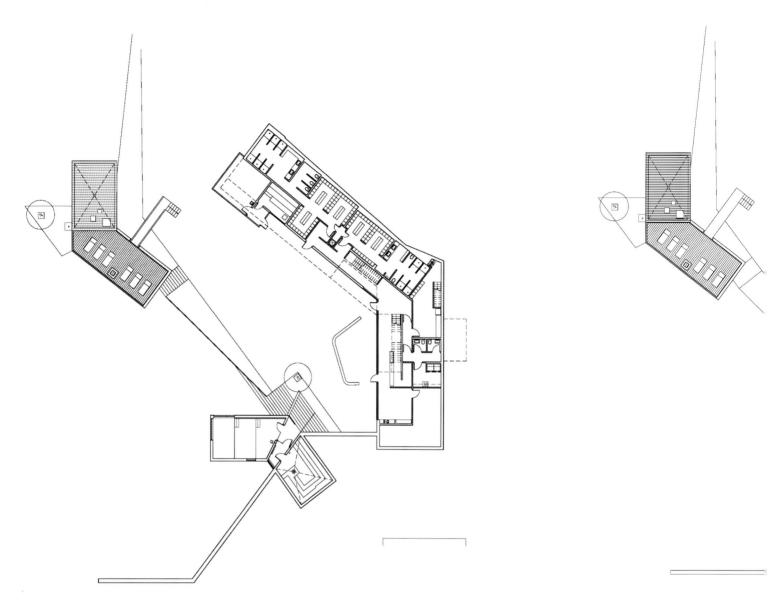

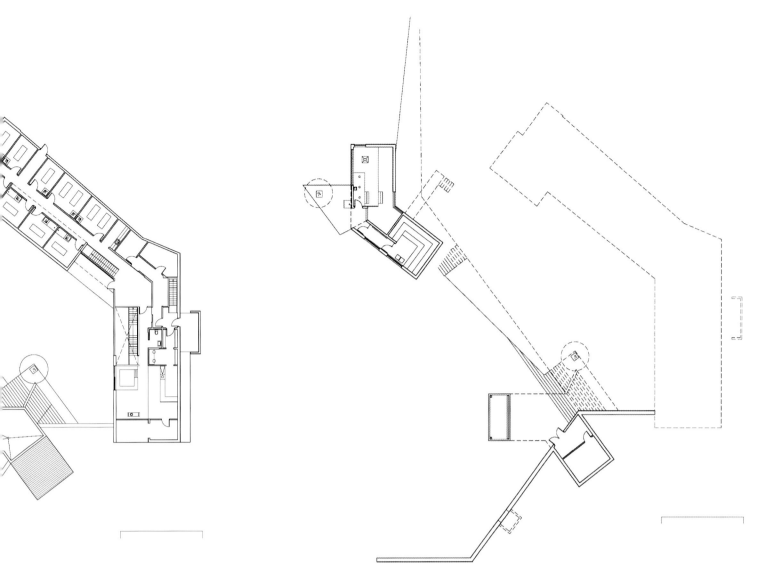

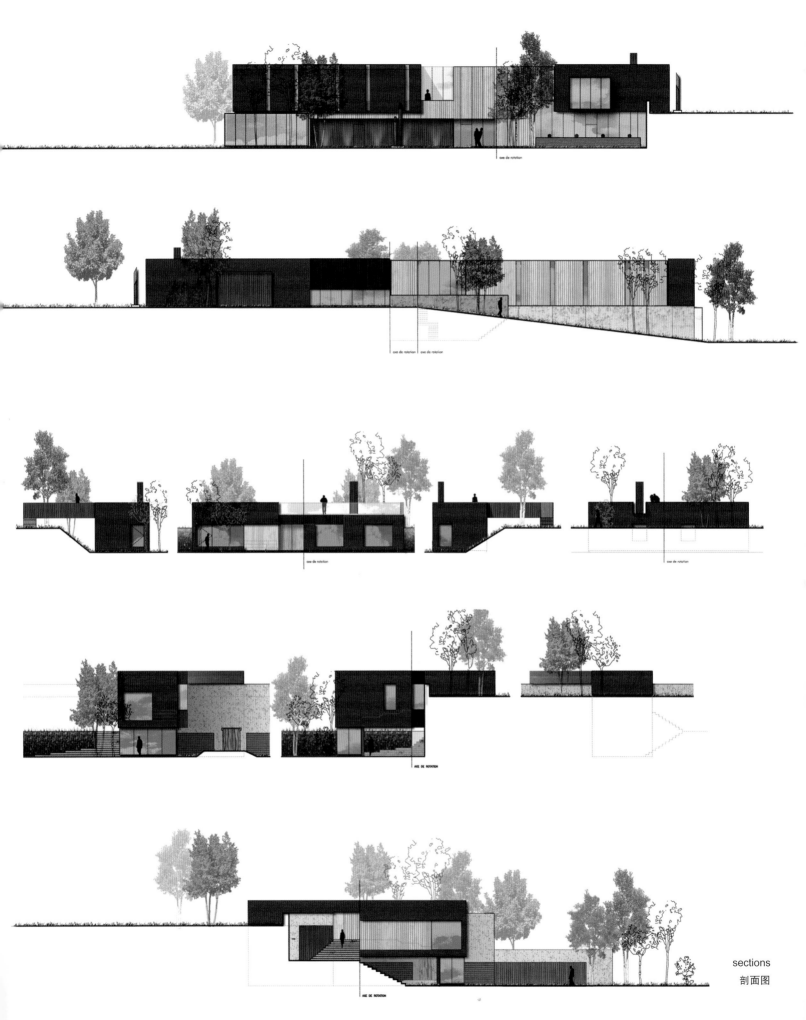

sections
剖面图

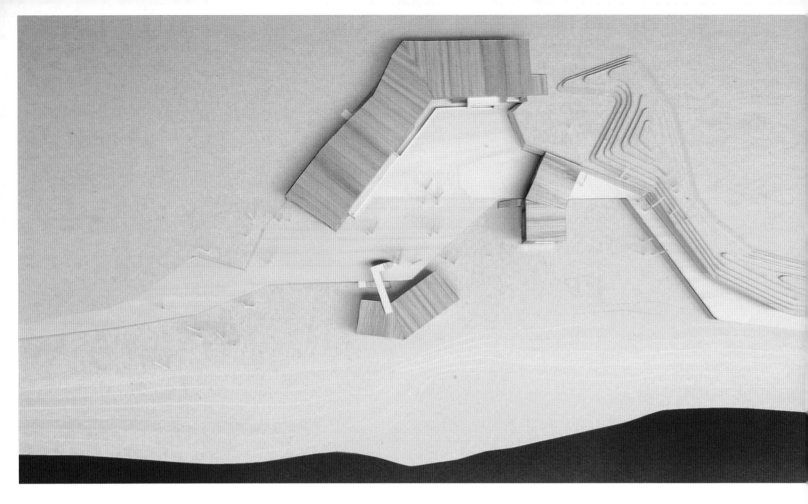
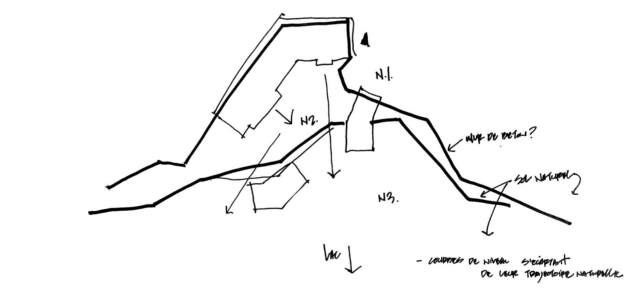
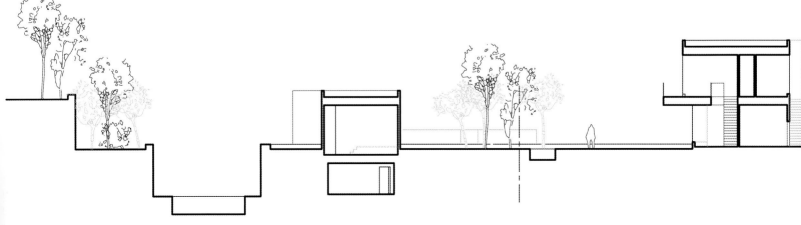

Architect: Cibinel Architects Ltd.

Landscape Architects: Hilderman Thomas, Frank Cram Landscape Architecture & Planning

Engineers: Crosier Kilgour & Partners Ltd, Epp Siepman Engineering Inc, Nova 3 Engineering Ltd, Williams Engineering Inc, M. Block & Associates Ltd.

Location: 12 – 19th Street North, Brandon, Manitoba, Canada

Area: about 2,000 m²

Photographer: Mike Karakas

BRANDON NO.1 FIREHALL

The dynamic new 2,800 m² Brandon Fire and Emergency Services Building validates the idea that a primarily utilitarian program, which often times results in a prefabricated solution, can become a sophisticated architectural project that contributes to its surrounding community and landscape while still fulfilling its demanding functional requirements and modest budget.

The facility is divided into two formal components, a fire hall wing and an administrative wing. The separation allows daylight to penetrate the buildings on all sides, and a single-loaded corridor animates the façade with human movement and activity. A minimally detailed, non-programmed transparent volume situated between the two wings acts as a dramatic entry into both sections of the facility, mediating the two programs with a thin hovering glulam bridge.

这个约2 800平方米的Brandon第一紧急救火中心表明一个以"功利主义"为价值导向的项目，在完成"服务于周边社区居民"的"功利性任务"的同时，也能够在有限的预算内，最大限度地履行其肩负的社会使命。

该救火中心由两部分构成，火警大厅和行政办公大厅，它们都呈"机翼"形状，好像一只张开翅膀、即将展翅飞翔的鸽子。机翼式的鲜明分区能够将自然光线均匀地引入这两栋建筑内。同时，它们又被一条互通的走道连接起来，活跃了双方的互动，有利于行政部门及时了解火警救护的信息，并且对其施救过程进行跟进。在翅膀中间设置了一扇透明的玻璃大门，作为共同的入口，又好像是连接它们的一个桥梁。

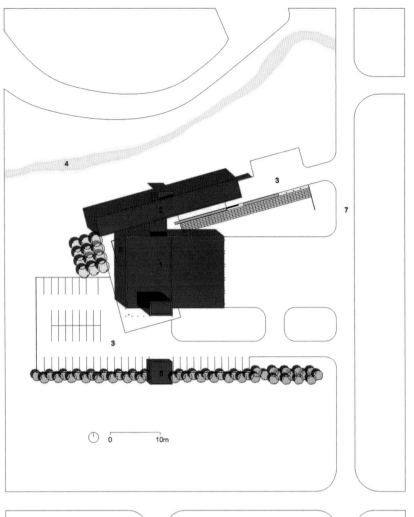

1. fire hall
2. administrative wing
3. parking
4. creek
5. outdoor space
6. cold storage
7. 19th street

site plan
总平面图

The public entry engages and welcomes the community with a generous landscaped public "plaza" enclosed by the museum to the north, and the apparatus floor to the south. The museum features a "Bickle", an 80-year-old fire truck situated as if it is ready to take off to the next call, while the apparatus floor houses the current emergency fleet. Extensive glazing surrounding the plaza highlights the rich history of the fire department in the city, and allows visitors a glance into a state of the art facility.

在公共入口处有一个社区交流广角，邀请社区居民前来集思广益，并且进行救火知识的普及宣传和演讲，同时允许居民参观火警救助人员平日训练的过程。该广角北面被博物馆包围，南面则连接着器械室。博物馆中最具特色的是一辆80岁高龄的救火卡车，它静静地站在那里，仿佛随时待命，准备出发；器械室内则陈列着各种救火飞行器。

1. E911 call centre
2. kitchen
3. store room
4. office
5. quiet room
6. classroom
7. glulam bridge
8. exterior deck
9. female locked room
10. male locked room
11. physical fitness room
12. study
13. captain room
14. sleeping quarters
15. hose tower
16. open to below
17. roof

second floor
二层平面图

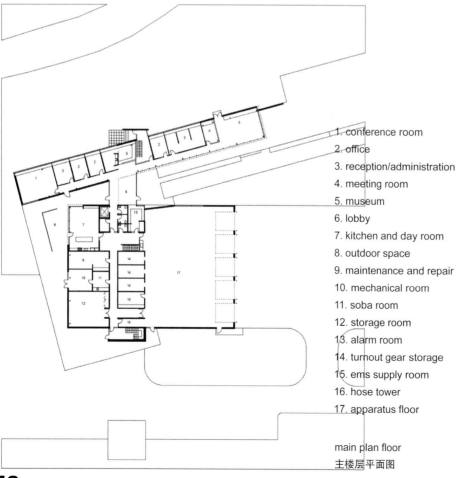

1. conference room
2. office
3. reception/administration
4. meeting room
5. museum
6. lobby
7. kitchen and day room
8. outdoor space
9. maintenance and repair
10. mechanical room
11. soba room
12. storage room
13. alarm room
14. turnout gear storage
15. ems supply room
16. hose tower
17. apparatus floor

main plan floor
主楼层平面图

A dramatic hose tower stands proud in the landscape and acts as an urban marker for people entering the city along the neighboring street. Impervious dark brick cladding descends the tower and wraps around the building's horizontal walking surfaces, allowing for a natural runoff of water to be reabsorbed by the site's indigenous grasses.

The solution integrates functionality seamlessly into the aesthetics, giving appropriate expression to the demands of the facility while creating unique and effective civic presence in the community.

一座高大的水带塔矗立在该中心院落的中央，成为该中心甚至整个城市的标志性建筑。它也是业务训练的所在地，并且与器械室和二楼的宿舍连为一体。具有不渗透性的黑色砖镶墙面从塔尖垂直下降，倾泻而下的清水仿佛一道道清澈的瀑布，成为院落中植被的灌溉水源。

Brandon第一紧急救火中心，集功能性和美学性于一体，创造性地履行着市政职责和社会使命。

Architect: Groupe Cardinal Hardy

LES LOFTS REDPATH

Les Lofts Redpath is located on the banks of the Lachine canal in Montreal. On a larger scale, the project constitutes a residential block composed of commercial spaces on the main floor. Though encountered by amid technical challenges, the developer still wanted to pursue the construction of residential lofts within these industrial brushlands. The current residents of the complex appreciate its specific industrial character and the exceptional attractions of the sector: the banks of the canal that have re-opened to nautical activities, amazing views of the mountain and downtown, a linear park, a bike path, etc.

Les Lofts Redpath工业群位于蒙特利尔Lachine运河沿岸。在很大程度上，该工业群被其周围众多的商业区所包围。尽管面临巨大的技术挑战，开发商依然想在这片密集的工业群中建造住宅区。该工业群所处的地理位置和自然环境得天独厚：运河沿岸向市民开放，可以自由地进行航海探险；登上高山，就可以俯瞰城市美景，一个呈直线状的公园，一条幽静的自行车道，等等。

The commercial spaces have distinct entrances that give directly onto the courtyard. The presence of commercial spaces also encourages contact with the canal's public spaces.

The courtyard constitutes a transition gateway as much for the residents as for the public. Its simple and elegant layout reinforces the architectural sobriety of the whole.

The courtyard is mainly composed of a grass garden, vines and shrubs, punctuated by willows suggesting the presence of water.

　　商业区有不同的出入口，直接通向住宅区庭院。商业区将住宅区和工业区自然地连接在一起，同时作为二者的屏障，也巧妙地保护了当地居民免受来自于工业区的污染。

　　庭院成为居民与户外大自然的天然过渡。其简洁的线条和高雅的风格，更加强调了建筑布局的整体连贯性。庭院里种着葡萄树和各种各样的花草、灌木，婀娜多姿的柳树，宛如少女弯腰时的妩媚，与其水中的倒影相映成趣。

The group of buildings serves as a complex of business, industrialization and residence. It complies with the sustainable principle, enhances work efficiency, and provides convenience to its residents and business circulation, embodying a significant model for the city architecture.

The project has been awarded the Finalist for the Awards of Excellence 2007 of the Ordre des Architects du Quebec, and represents a cultural interest recognized by Park Council of Canada and the City of Montreal.

该工业群集商业、工业、住宅于一体，在遵守环保原则的同时，提高了工作效率，方便了居民生活，加速了商业流通，堪称城市建筑的典范。

该设计荣获2007年度魁北克建筑设计师最佳杰出奖，并且已经被加拿大公园协会和蒙特利尔市政府列为"文化景区"，成为城市中独特的一景。

Architect: KPMB

MANITOBA HYDRO PLACE

Manitoba Hydro Place This is a world-class office tower which is a model for the next generation of extreme climate responsive architecture. Designed by the integrated design consortium of Kuwabara Payne McKenna Blumberg Architects (Toronto), Smith Carter Architects (Winnipeg), Transsolar Klima Engineering (Stuttgart), the tower has already gained attention with the prestigious "Best Tall Building in North America" award that is granted by the CTBUH (Council for Tall Buildings), the world's leading body dedicated to the field of tall buildings and urban habitat. Already generating international interest, the project has appeared in the Princeton University Press, the Architectural Press, and various other journals in Europe and Asia.

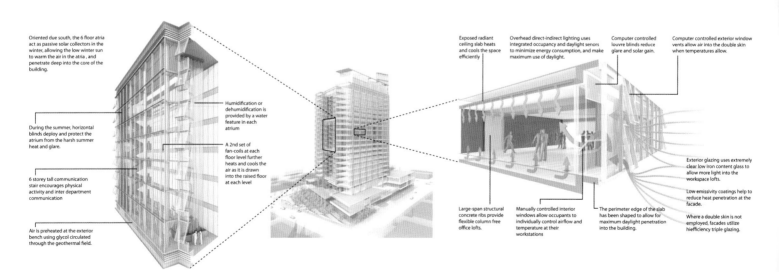

Winnipeg, located in the geographic centre of North America is one of the coldest large cities in the world. From the start, in 2003 the client, Manitoba Hydro, established ambitious goals and mandated the project be conducted within a formal Integrated Design Process (IDP). It was to be designed to optimize passive free energy and 100% fresh air year round in an extreme climate that fluctuates from -35°c to 34°c and without compromise the comfort of 2,000 employees. The design integrates tested environmental concepts in conjunction with advance technologies to achieve a "living building" that dynamically optimizes its local climate.

这一世界级的Manitoba Hydro商业办公大厦在应对极端气候条件方面，堪称新一代建筑群中的经典之作。该设计由多伦多Kuwabara Payne McKenna Blumberg建筑设计院、温尼伯Smith Carter 建筑设计院和斯图加特Transsolar KlimaEngineering工程局组成的联合财团策划并实施，并荣获了由CTBUH（致力于高大型建筑以及都市住宅区研究的世界顶级权威机构）颁发的"北美最佳高大型建筑奖"。此外，普林斯顿出版社、建筑出版社以及欧亚地区的众多期刊社都对其进行了专门性报道。

Manitoba Hydro大厦位于温尼伯，处在北美地理的中心位置，是世界上气候最为寒冷的大城市之一。从2003年开始，该大厦就制定了IDP（整合式设计）建筑目标，在极端的气候条件下，实现"被动的自由能源"和100%新鲜空气的全面、充分、不间断利用，使温度

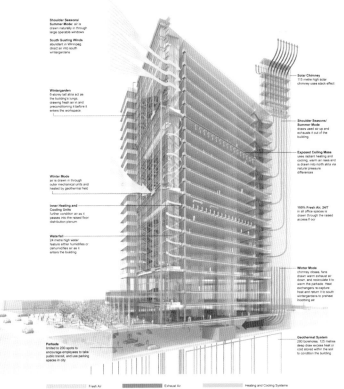

At the same time, it was to play a major role in the revitalization of the downtown, demonstrate architectural excellence, and most importantly demonstrate the client's recognition that its employees are its greatest asset by delivering a highly supportive, healthy workplace environment.

在-35℃至34℃之间自由波动，并且不对其内部2000名员工的身体健康造成任何消极影响。该设计践行"打造富有生命力的环境建筑"这一新型设计理念，充分利用当地不利的气候条件，求得"最佳化生存"，将"消极"变为"积极"。同时，作为该地区最出色的建筑之一，Manitoba Hydro大厦重新点燃了温尼伯市中心的生命力。更为重要的是，它为在这里工作的员工营造了一个健康、高效、舒适宜人的工作环境。

Architects: Fournier, Gersovitz, Moss, Drolet et Associates Architects
Location: Hemmingford, Quebec, Canada
Area: 2,400 m²

CIDRERIE LA FACE CACHÉE DE LA POMME

Cidrerie la Face Cachée de la Pomme is a complex of argricultural production, family residence and travel.

In order to perfectly achieve these three functions, the project has three phases: architecture, landscape architecture and interior design. The architects respect the Scottish-inspired vernacular architecture, merging it with a resolutely modern, contemporary architectural language. Inspired by farms in southern Italy, the architects designed the residences and farm facilities to create a unity that is both aesthetic and functional. The yard can be a dock for unloading merchandise or be transformed into a tasting area for visitors. The grouping blends harmoniously into the landscape, enabling it to be a paradise for relaxing.

Sustainable development is part of the framework of the project, integral to both the facilities and the production processes. Particular attention was paid to the lighting; natural light was emphasized, which is particularly important for architectures under the cold climate. Therefore, Cidrerie la Face Cachée de la Pomme is located to the orientation which is towards plenty of natural light.

Cidrerie la Face Cachée de la Pomme农场别墅是集农业生产、家庭居住,以及旅游于一体的综合别墅。

为了能够更好地实现这三项功能,该工程从建筑设计、景观设计和室内设计三个方面来对Cidrerie la Face Cachée de la Pomme农场别墅进行整修。建筑师借鉴苏格兰建筑的设计风格和建筑语汇,并将意大利南部农场的建筑模式引入其中,功能齐全且造型美观,真可谓"一箭双雕"。宽敞的庭院被用来整齐地摆放各种农具,好像一座码头。这里与周围的自然景观浑然一体,是休闲度假的绝佳圣地。

可持续发展的设计原则贯穿该工程的始终。值得一提的是,建筑师特别强调自然采光,这在寒冷的自然条件下尤为重要。因此,这里的建筑均建在自然光线足够充足的地方。

Architects: Brière, Gilbert + Associés Architectes
Location: Québec, Canada

REFURBISHMENT AND EXTENSION OF THE CRÉVHSL HEADQUARTERS

The project is defined in two parts:

The refurbishment of the existing building which accommodates several offices and the main reception and A contemporary extension to the rear, where the new board room and the employee's cafeteria is located.

The refurbishment included mainly interior renovation and restoring to the original states the existing exterior doors and windows, and repairing the masonry and the roof.

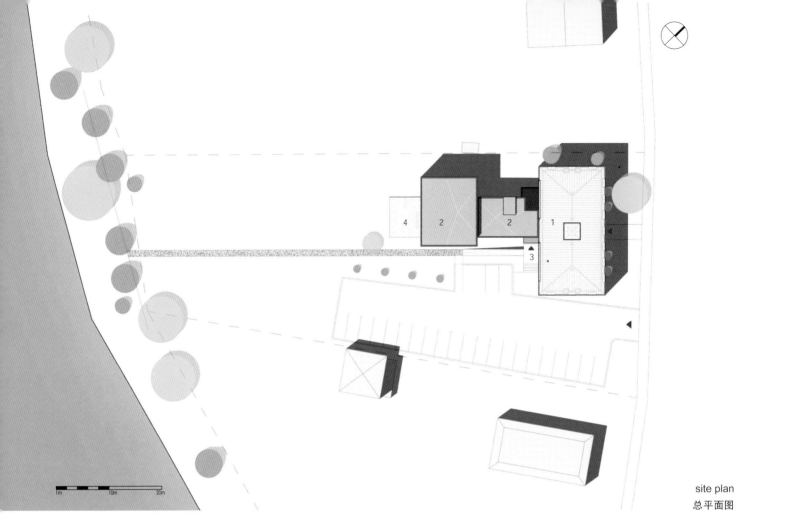

site plan
总平面图

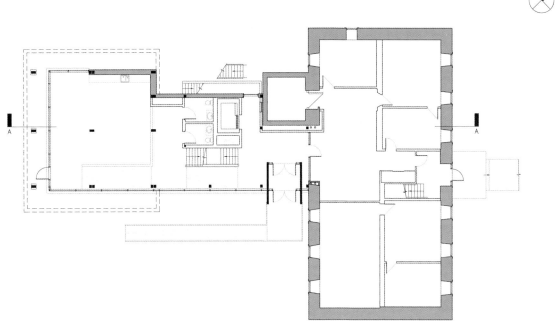

first floor plan
一层平面图

Three conceptual gestures helped to achieve these goals successfully:

First, the off-axis footprint, from the existing building of the extension, allows visual openings towards the St Lawrence river from both the new and existing building; Secondly, the "suspended" wooden volume, which wraps the new board room, provides an adequate response to the massive stone volume of the existing building; Thirdly, the integration of two atriums creates open spaces to accentuate the suspended wooden volume from the inside and opens views towards the river and site from the new and existing building.

The project includes sustainable development measures. The use of eastern Quebec cedar and wood for the exterior siding aims to enhance this natural resource of Québec. In addition, the project uses geothermal energy for heating, air conditioning, linked to heat recovery.

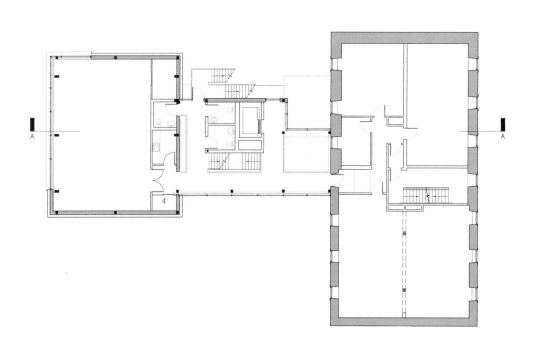

second floor plan
二层平面图

Crévhsl Headquarters整修工程包括：

整修包含办公室和主要接待大厅的现有建筑，在现有建筑的后部适当地予以延伸，在那里，搭建董事会专用房间和员工咖啡屋。

这次整修包括对室内空间的粉刷和外部门窗，以及屋顶砖石的重新加固。这次整修主要达到以下三个目标：1.对该建筑后部的延伸，使其与附近的St. Lawrence River连成一体。这样，该建筑紧邻河流，风景秀美。2.董事会专用房间的室内材质均使用当地的木材，与以砖为主要材质的整栋建筑形成鲜明的对比。3.董事会专用房间的两个天台是房间与户外的天然衔接，使房间开阔且明亮。

该建筑整体上遵循了可持续发展的设计原则，就地取材，充分利用当地的木材和雪木松，并且利用地热能供暖，既环保又高效。

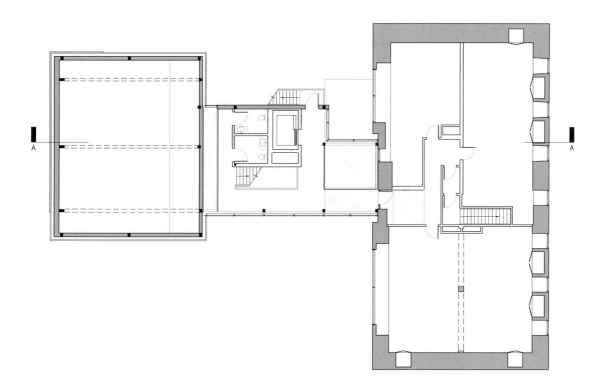

third floor plan
三层平面图

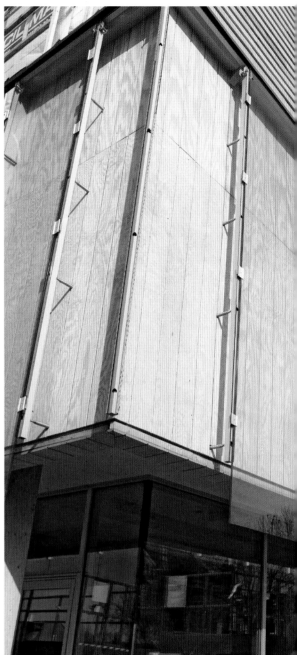
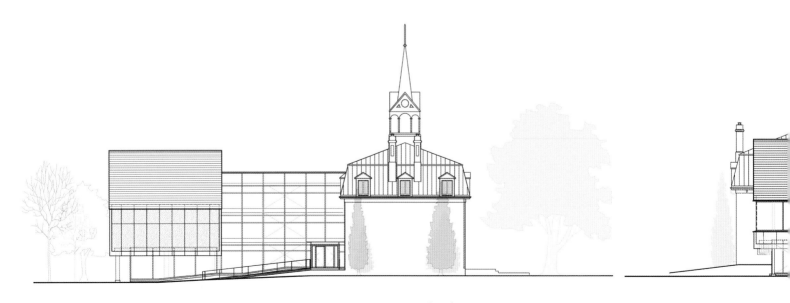

elevation
立面图

196

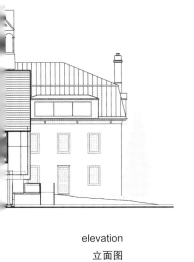

elevation
立面图

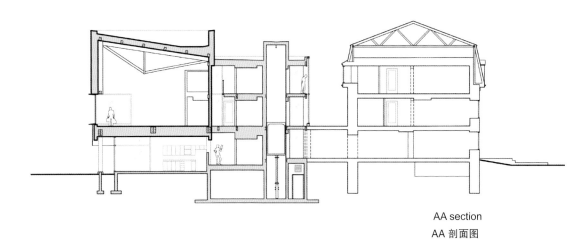

AA section
AA 剖面图

197

Architects: Blouin Tardif Architecture-Environnement
Location: Charltvoix, Quebec, Canada
Photographer: Stéphane Groleau

STATION BLÜ

This resort complex situated on the outskirts of the Charlevoix region includes a restaurant, steam bath, sauna, massage areas, hot and cold pools, and many relaxation spaces. These functions are shared among three pavilions articulated around a landscaped area bordering the river.

To unify the project grouping, the buildings' envelope is a black-coloured wood covering. The various grades used give a more random texture, referring to vernacular structures.

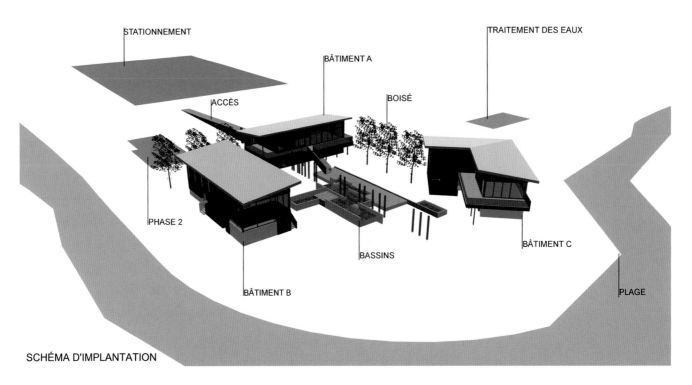

SCHÉMA D'IMPLANTATION

　　该度假胜地坐落于临近Charlevoiv地区的郊区，包括餐厅、温泉、桑拿、按摩室、游泳池，以及许多娱乐场所。三座精致的小凉亭屹立其中，为这里增添了无尽的情趣。

　　为了环保，这里的建筑均使用取材于当地的黑色木材。该黑色木材散发着浓郁的乡土气息。游客通过东面的一条长长的走道进入这里后，首先映入眼帘的是高高的天花板和彩色的窗户，站在窗前，就可以尽情欣赏窗外迷人的风景。

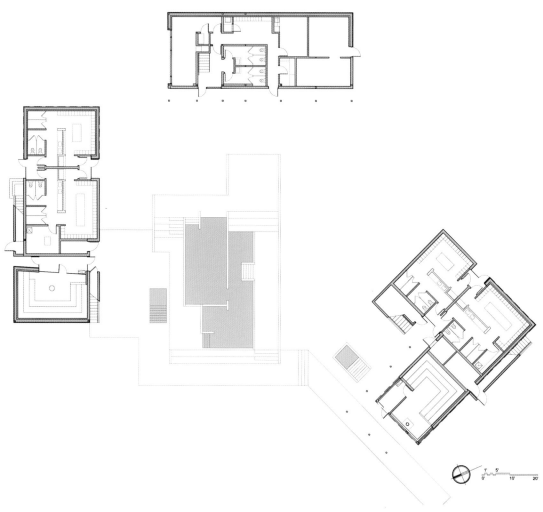

first floor plan
一层平面图

Once visitors enter the volume, the high ceilings and picture window project them into the exterior landscaping and the countryside.

A walkway is cut through the east wall of the main building; this is the only distinctive element marking the entrance to the project.

The interior spaces are uncluttered, giving the natural landscape the starring role. Bright colours are used only for specific elements, such as the restaurant's large banquette and the steam bath.

This resort has been widely considered as an excellent example of integrating well with its surroundings.

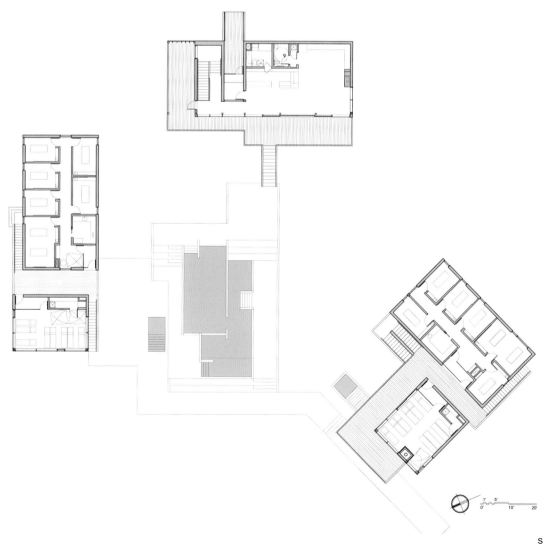

second floor plan
二层平面图

室内空间的外围并没有增设围栏，目的是能够使室内外空间浑然一体。

该建筑的色彩整体而言，相对素雅，然而，大型宴会厅和桑拿室则使用了比较鲜明的色彩。

该度假胜地堪称与其周围景观完美融合的经典建筑。

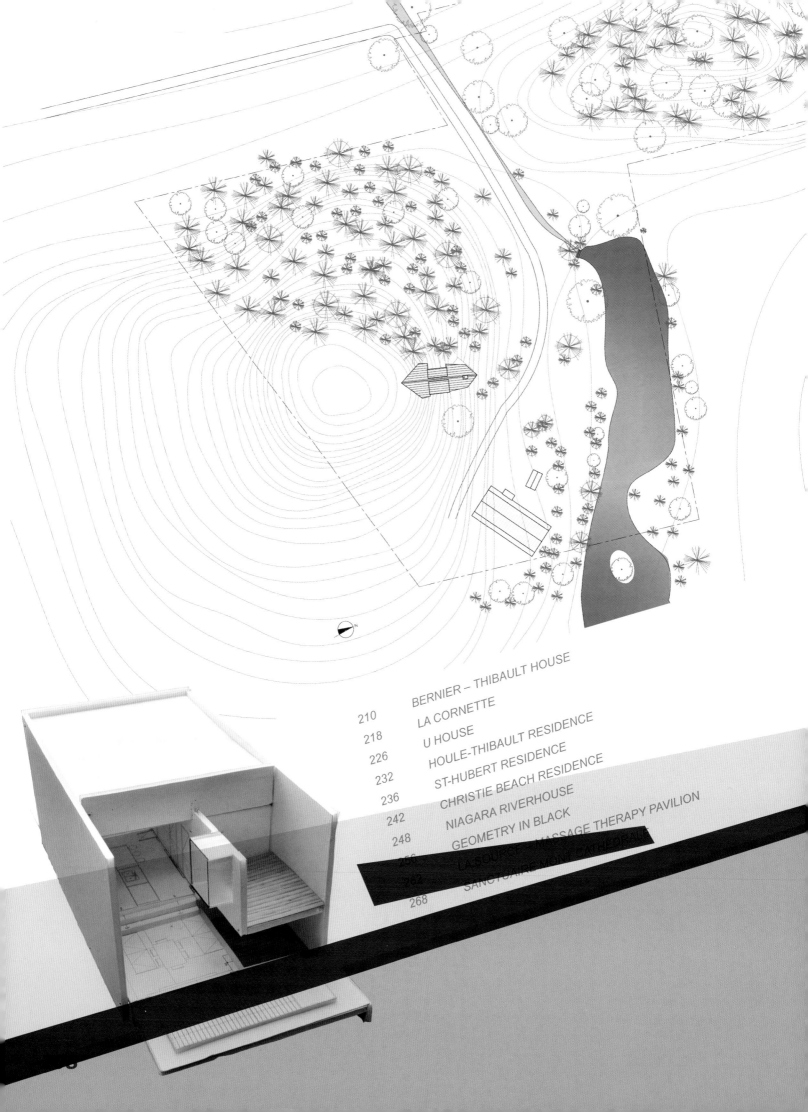

	BERNIER – THIBAULT HOUSE
210	LA CORNETTE
218	U HOUSE
226	HOULE-THIBAULT RESIDENCE
232	ST-HUBERT RESIDENCE
236	CHRISTIE BEACH RESIDENCE
242	NIAGARA RIVERHOUSE
248	GEOMETRY IN BLACK
256	LA SOURCE – MASSAGE THERAPY PAVILION
262	SANCTUAIRE MONT CATHÉDRAL
268	

RESIDENCE

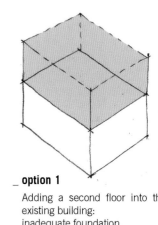

_ option 1

Adding a second floor into the existing building:
inadequate foundation

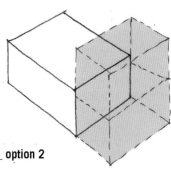

_ option 2

Two storey extension into the backyard:
forbidden by the Municipality

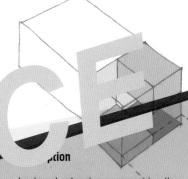

_ option

backyard extension compressing the spaces on numerous split-levels

Architect: Paul Bernier Architect
Location: Montreal, Quebec, Canada
Photographers: Marc Cramer, Paul Bernier, Vittorio Viera

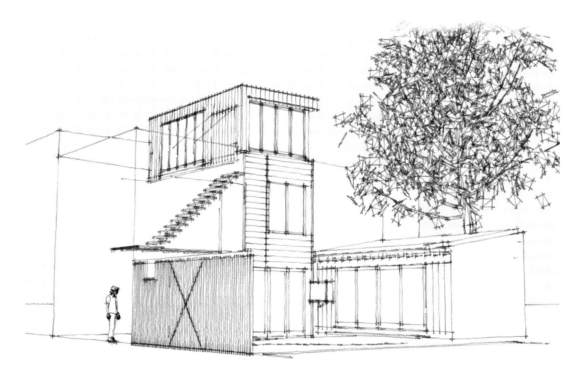

BERNIER – THIBAULT HOUSE

This extension and transformation of the house has to be made to allow for a family of four. Two rooms have to be added, one for the kids and one for the adults. Besides, the designers extend the house and preserve the quality of the garden while working to increase natural light. The house aims to become a perfect integral of the families' spirit and materiality.

Two boxes made of glass and wood, simple volumes of similar dimensions, are added to the original house. One box is placed on the roof and the other one in the garden under the big maple. A vertical slice of the original garden side wall was taken out and replaced by a wood structure wall that allows for openings on the garden and that acts as a formal link between the two boxes.

该住宅位于临近蒙特利尔Royal山脉的高原地带。

两个由玻璃和木头制成的盒子，具有相似的尺寸和容积，被创造性地融入整体设计。一个被放在屋顶，另一个则被放在花园中枫树的下方。原来花园中垂直的墙面被拆除，被木质墙面所代替，作为开启花园的大门，并且充当两个盒子之间天然的连接物。

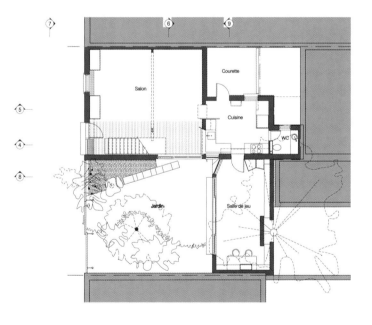

first floor plan
一层平面图

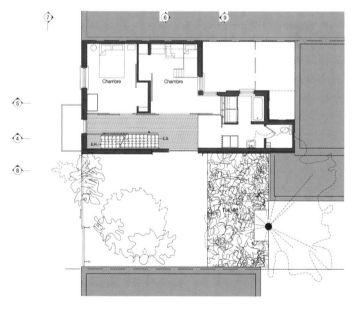

second floor plan
二层平面图

The box on the roof shelters the adult room, like a big hat. Standing on it, people can see the city and the sunrise. The box also smoothly leads the natural light into the inside.

The box in the garden, a playroom for the children, is connected to the interior living spaces and opens up on the courtyard with wide glass doors as a pavilion in a garden, bright and transparent.

In addition, the ground floor, which serves as the space for family life, is an L-shape that, in the three sides, wraps around the garden and opens to the outside on one side, becoming an extra room in the summer, without any depressed feelings or darkness.

屋顶的盒子仿佛一顶大帽子,精心呵护着父母房。登上屋顶,透过盒子,就可以看到日出,城市的美景。同时,盒子将自然光线顺理成章地引入室内,使室内充分享受自然光照的同时,也不至于暴晒。

花园中的盒子,是孩子玩耍的小屋,和起居室连成一体,就仿佛花园中矗立着一个透明的小凉亭,晶莹剔透。

此外,地下室成L形,三面紧紧地将花园包围,另一面则敞开通风,成为盛夏时节中的又一乘凉之地,毫无憋闷、阴暗之感。

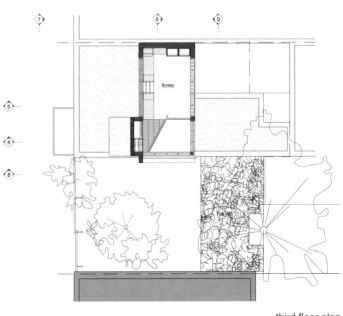

third floor plan
三层平面图

Architect: YH2-Yiaconvalicis Hamelin Architects
Builder: Emmanuel Yiacouvakis
Project Team: Marie-Claude Hamelin, Loukas Yiacouvakis
Location: Township of Cleveland, Quebec, Canada
Area: about 280 m²

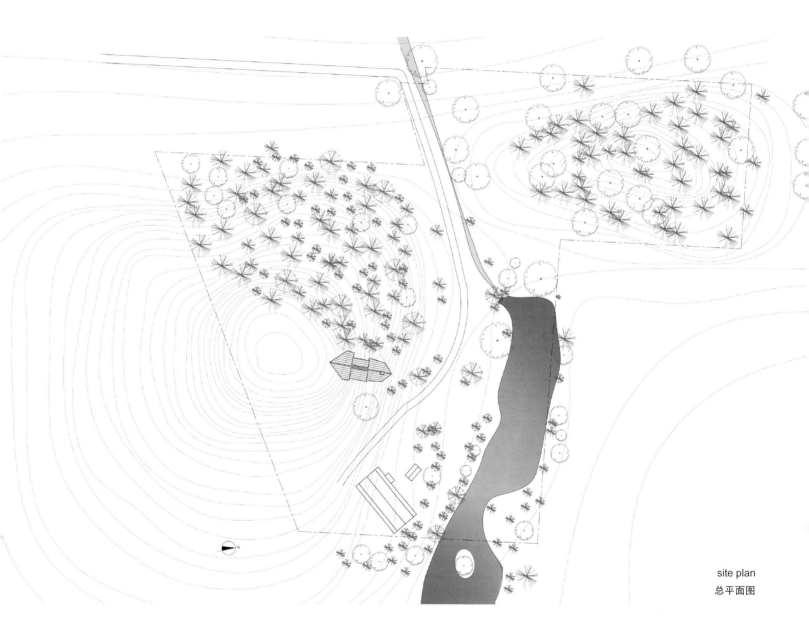

site plan
总平面图

LA CORNETTE

Built on the slope of a small hill, La Cornette is a country house open to the pastoral landscape that surrounds it. This house for celebrations and holidays, designed for two families, is set into the naturally uneven terrain in a way that brings each level into direct contact with the surrounding natural environment. It offers a resting place for all guests under its large gable in a series of bedrooms and unusual sleeping areas.

An out-scaled structure, like the agricultural buildings that surround it, the house is both traditional in its morphology and innovative in its use of materials. Shingled with raw fibre-cement panels on the walls and roof, it is a house beyond the domestic scale, simple and rot-proof, capable of standing the test of time. The house is striated with bands of horizontal windows, giant louvers that cut the sun at its most powerful, with new points of view at each level. It is protected by its wimple from the hot summer sun and inundated with light in the winter, needing neither air-conditioning nor heating on sunny days.

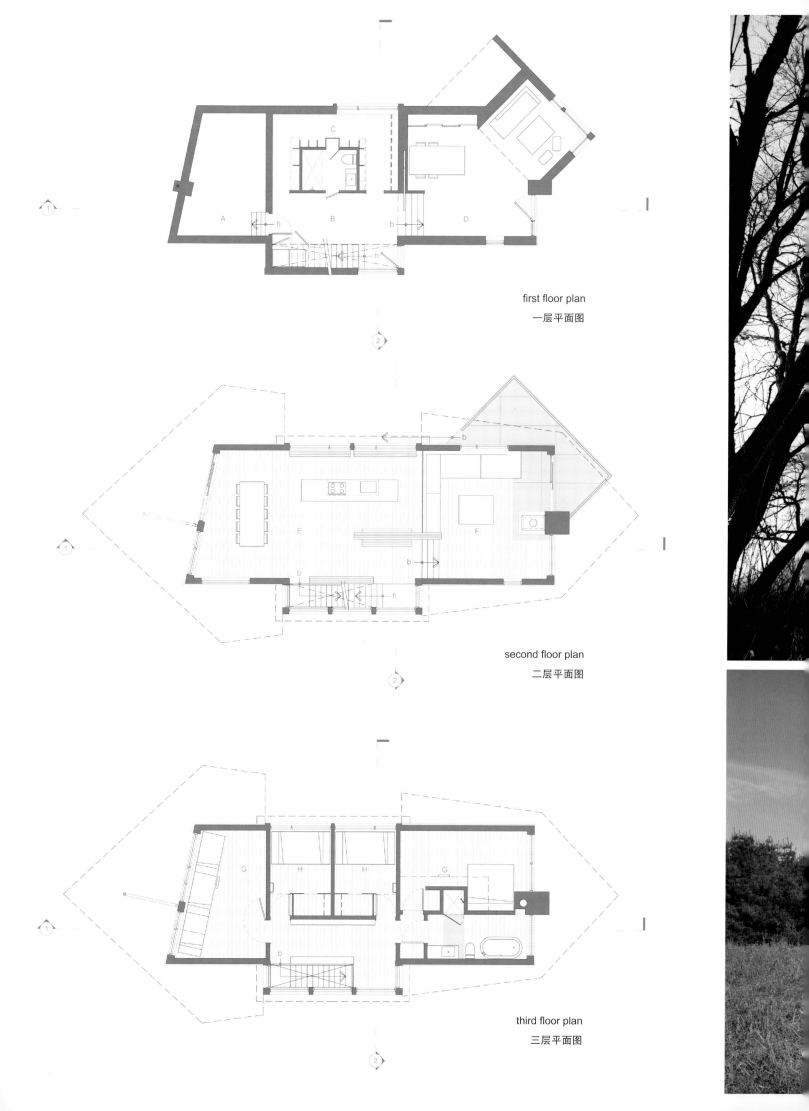

first floor plan

一层平面图

second floor plan

二层平面图

third floor plan

三层平面图

east elevation
东立面图

south elevation
南立面图

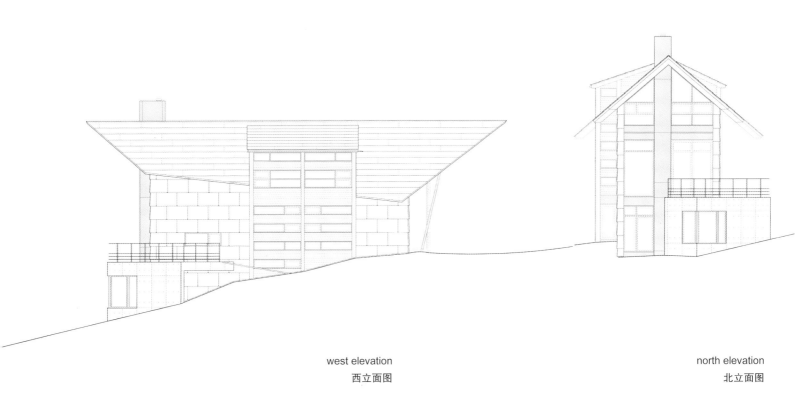

west elevation
西立面图

north elevation
北立面图

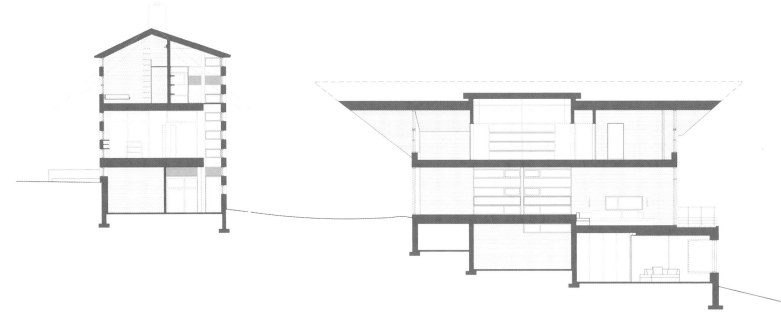

sections
剖面图

224

The interior is in wood, painted or natural, in planks or panels, composed almost exclusively of made-to-measure furniture pieces, such as the refectory table for meals, the large wrap-around couch, the balustrade bookshelf along the stairway, the night-lights made of aluminum panels with cut-outs of fireflies, fish, and frogs; comfortable beds.

It is a playground for architects, children and adults, as well as a vacation colony lost in the countryside.

La Cornette乡村度假别墅建在一座小山的山坡上，被其周围迷人的田园美景所环绕。它能够供两个家庭来此庆祝节日或者度假。独特的三角墙面将一系列卧室和休闲空间紧紧包围。每一处景致的细节设计，都旨在与当地的地形条件形成完美的和谐。

就像其周围原生态的草屋一样，该别墅在其外形和使用材料方面都非常传统，"在传统中追求创新"。墙面和屋顶均由纤维水泥平板搭建而成，线条简单，防腐蚀，完全能够经得起时间的考验。别墅的每一扇巨大的百叶窗呈水平状，在冬日里，"慷慨地"将窗外的自然光线反射进来，强劲有力；每一扇百叶窗的外部边缘上，都镶嵌着一条美丽的窗帘，从远处看，悬挂着窗帘的百叶窗，好像夏日里的一件"蓑衣"，有效地阻挡了炎炎烈日般的炙烤。在这里，冬天不需要供暖，夏天不需要打开空调。

室内摆设着木质家具，使用与户外自然相一致的颜色粉刷，或者干脆保留其原来的色调。各种定制式家具，例如，食堂式宽大的餐桌，松软的沙发，沿楼梯而建带有扶手的书架，萤火虫、金鱼、青蛙形状的夜灯，以及舒适的大床，令空间尽显温馨与整洁。

这里是设计师的竞技场，这里是孩子们的游乐场，这里是成年人休养生息的世外桃源。曾经失去的田园美景，曾经失去的天堂，一切都从这里，重新开始。

Architect: Natalie Dionne
Photographer: Marc Cramer

U HOUSE

Situated near a railway line on the outskirts of a trendy Montreal neighbourhood, the design of U House responds to a harsh urban context, respectfully shutting out the outside world by concentrating services around the periphery and creatively keeping away from the noisy outside world. Constructed with the precision of fine cabinetry, the articulated panels of wood, glass and steel that line the courtyard open and close to invite the outside in and the inside out.

U House坐落于临近蒙特利尔的一个时髦郊区的火车沿线上。其设计是对日益严峻的都市环境污染这一问题的有力回应，它将种类齐全的配套设施紧密地安置在该地区的周边，创造性地远离了嘈杂的外界环境，仿佛一个"世外桃源"。设计精致的手工家具独具特色，矗立在庭院门口的由木头、玻璃和钢铁制成的镜板，魔幻般地一开一闭，使院内的美景与外面的自然世界融为一体。

U House was designed by architect Natalie Dionne to house both the family residence and her office space. She exploits the idea of fluidity between interior and exterior spaces. The Ipê wood deck which is an extension of the dining room's inlaid wood floor helps to confound the boundary between inside and outside when the large accordion style door is opened. At the other end of the L-shaped deck, one of the original building's huge French windows is actually a garage door that also opens up fully to seamlessly integrate the living room with the garden. For added spatial continuity and a layered effect, the wood and steel façade elements of the new additions were mapped to the interior.

Doors and windows of Cedro wood, marine plywood panels and cedar bifold shutters are all stained for uniformity and mimic the color of the brick on the original building, creating a brand-new look.

The shutters upstairs are opened. This ensures sufficient ventilation, guarantees privacy, protects from the rain, and keeps the south-facing rooms nicely cool. What deserves to mention is that in summer, at night, whether it is raining heavily outside or not, it is always quiet and comfortable. What a pleasant resort!

该设计由建筑师Natalie Dionne完成，兼具家庭居住和办公空间的双重功能。她将"实现室内和室外的自然、和谐的流动"作为设计的亮点。当手风琴形状的大门敞开时，延伸出来的餐厅木质地板，室内和室外神话般地"浑然一体"了。在"L"形的地板的另一端，室内巨大的落地窗好像一道大门，坚定地促成起居室和花园的"亲切握手"。为了增强空间的连续性和分层效果，木质和钢板的立面元素被镶嵌到室内。

由Cedro木材制成的门窗、海洋胶合板和雪松百叶窗板避免了建筑材料使用材质上的千篇一律，丰富了建筑外观的色彩，使整个建筑焕然一新。

打开楼上的百叶窗，既保证了足够的通风，同时也使空间具有了良好的私密性。特别值得一提的是，在盛夏的夜晚，无论外面下着多大的雨，这里依然静谧、安逸。南北朝向的房间，凉爽宜人。此情此景别有一番风情！

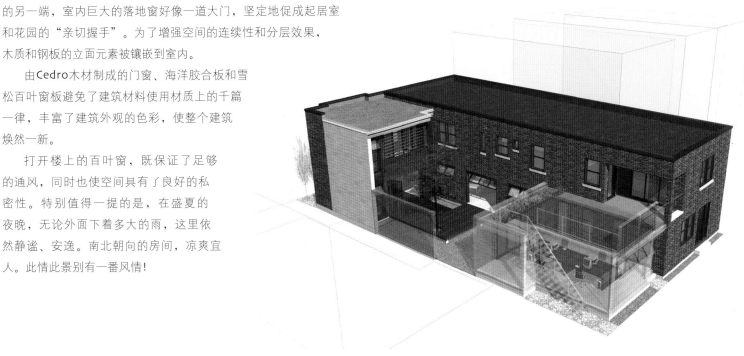

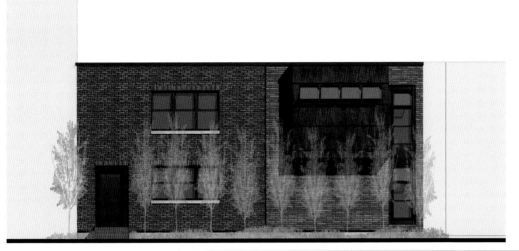

Architect: Chevalier Morales Architect

Project Managers: Sergio Morales, Stephan Chevalier

Project Team: Sergio Morales, Stephan Chevalier, Karine Dieujuste, Christine Giguère, Samantha Hayes

Location: Mont-Tremblant, Quebec, Canada

Site Area: 4,500 m²

Built Area: 300 m²

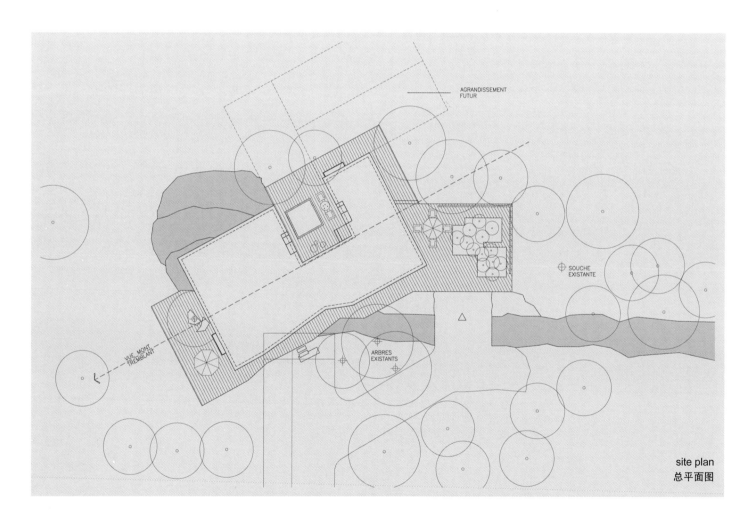

site plan
总平面图

HOULE-THIBAULT RESIDENCE

Built on a mountainside with incredible views of Tremblant Montain, the Houle-Thibanlt Residerce embodies a sense of obvious nostalgia of traditional architecture, as well as tastes old and exquisite qualities in natural areas.

Following the client's initial request to have a timber-framed house, the designers worked with the help of a local artisan and after multiple iterations of the frame, they ended up not only achieving the simplest possible shape but we also managed to dissimulate all anchorage so the frame could be read without the interference of steel plates and bolts.

The residence sits silently on a rock vein, as if it had emerged naturally from the landscape. The mineral base contains the garage, storage and technical spaces

When entering the residence, in a double-height space, a set of four large windows allows an immediate view of the top of the trees. On both south and west sides, the shape retracts itself on the ground floor, creating an integrated pare-soleil that helps maintain these windowed spaces shaded during the summer.

The residence is full of recreational and natural fragrance, small and lovely.

Houle-Thibanlt木质小屋坐落于Tremblant山的半山腰上。其怀旧式的建筑风格，勾起了来此旅行的游客和驻足观光者强烈的怀旧情绪，在大自然中，细细品味古老、精致的韵味。

为了满足业主对"木质小屋"的渴求，建筑师特意请来了一位当地著名的工匠，经过对纷繁复杂的木料进行仔细筛选，简洁、明了的造型诞生了，同时，框架锚被巧妙地掩饰起来，使木质小屋能够自由地站立，不受钢板和螺栓的束缚。

木质小屋静静地站在一个大岩石上面，好像突然从周边的自然景观之中"冒"出来，别有一番风情。岩石的矿物质构成了小屋的地下基板，车库、储藏室和技术空间都在这里。

进入小屋，双层空间中，4扇高大的落地窗和窗外的绿树一样高。在南面和西面，小屋的形状呈缩进状，在盛夏酷暑之时，有效地防止了室内不受户外炎炎烈日的炙烤。

该木质小屋集娱乐性与自然性于一体，娇小、可爱。

Architect: Naturehumaine

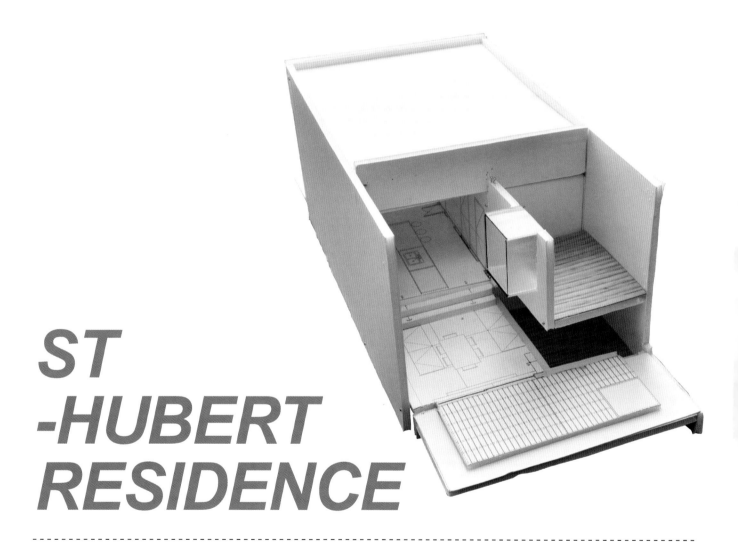

ST -HUBERT RESIDENCE

The clients wished to enlarge their 74 m² bungalow by adding a second floor to the existing structure. However, the poor conditions of the foundations quickly proved this option impossible. In turn, the architects studied the possibility to build an extension into the backyard. Two major constraints were to be found: 1. the Municipality forbades constructing higher than the existing roof membrane, 2. the presence of the 1-meter rock into the ground made the construction of a basement very costly. From those limitations, an unconventional and affordable solution was developed: compressing the spaces on numerous split-levels to yield the desired rooms, with a stunning double-height dining room and a generous provision of natural light.

业主希望将这个74平方米的别墅扩大,在现有的基础上,增加一层的空间。但是,现实条件很快就证明这个方案并不可行:1. 现有的别墅高度已经是允许范围内的最大高度,再次增加高度则将不会被市政府批准;2. 矗立在庭院里的1米的岩石令建造一个地下储藏室极为困难。于是,建筑师采用了一个超常规和成本合理的方案:将现在每一层的空间压缩,在匀出来的空间中建造更多的房间,并且打造一个具有双倍高度的餐厅。通过餐厅,将窗外的自然光线顺利地引入室内。

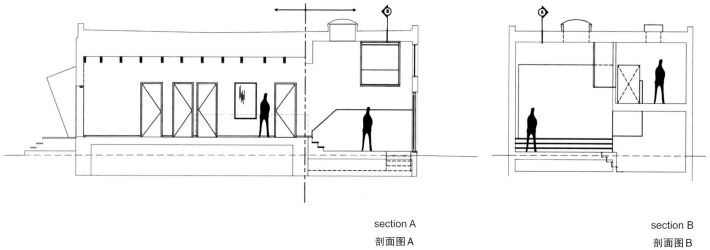

section A

剖面图 A

section B

剖面图 B

_ option 1

Adding a second floor into the existing building: inadequate foundation

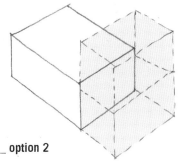

_ option 2

Two storey extension into the backyard: forbidden by the Municipality

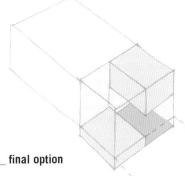

_ final option

backyard extension compressing the spaces on numerous split-levels

The first gesture was to lower the new dining room to the level of the exterior terrace and to link it with the kitchen and music room through a vast open space. Suspended atop the dining room, a translucent reading cube emerges from the master bedroom, creating a quiet and comfortable reading space. The kitchen is organized around an over-dimensioned central counter that becomes the focus of the social life in the house.

To meet the extremely tight budget, the selected building materials were deliberately left raw and untouched. The floor is covered mostly with an antic waxed maple flooring while the dining room uses fibrocement panels. The roof structure of the existing house was left exposed and painted in white to lighten up the spaces. The exterior back facade is covered with black pine planks with industrial corrugated steel sheet with galvalum finish.

Playing in a subversive manner with the numerous constraints, the architects yielded a unique project. Simple and modest, St-Hubert residence offers nonetheless a rich spatial experience with generous and luminous spaces.

第一个步骤是使餐厅和窗外露台的高度相一致，以厨房和音乐室作为两者的自然衔接，这样，一眼望去，露台、餐厅、厨房、音乐室，空间开阔了许多，自然光线照到这里，一气呵成。在餐厅通往主卧那并不宽敞的空间之中，建立一个读书角。这样，在原本狭小的空间里，还能够为业主打造一个宁静的读书小天地，不受外界的干扰。厨房处于整栋别墅布局的正中央，且占据的空间最多，因为，俗话说"民以食为天"嘛。

为了不超出本来就比较紧张的预算，建筑师精挑细选，选择的材料皆为原生态或者特别实用的。例如，除了厨房，其他空间均以蜡枫材料铺设地板，既防水又美观，其淡棕色为空间注入了一丝淡雅的气息；厨房则使用石棉水泥铺设而成；屋顶保持原生态，未作任何修饰，仅仅是将其粉刷成白色；建筑后背面由抛光的工业褶皱松板拼搭而成。这样，整栋别墅从外观上看起来十分整洁、透亮。

经过改造后的别墅，线条简洁，明亮开阔，成本低廉，洁净环保。

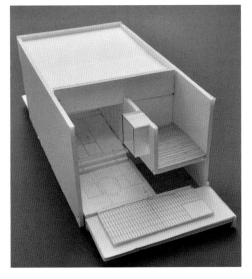

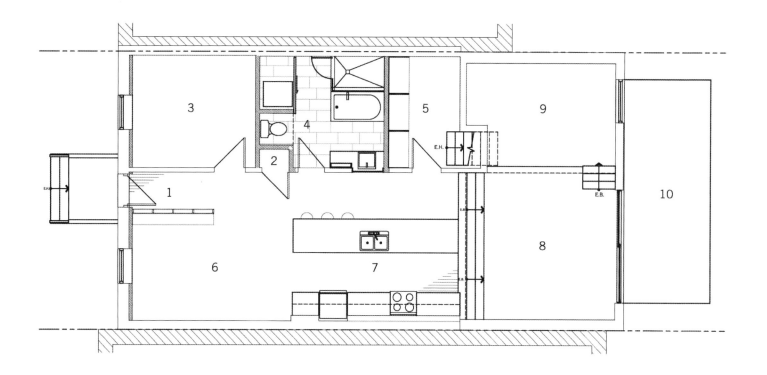

1. hall	3. bedroom	5. walk-in	7. kitchen	9. living room	first floor plan
2. cloakroom	4. bathroom	6. music room	8. dinning room	10. exterior terrace	一层平面图

Architect: Altius Architecture Inc.
Leading Design Architect: Graham Smith
Location: Christie Beach, Ontario, Canada
Area: 557 m²

CHRISTIE BEACH RESIDENCE

Set on the south shore of Georgian Bay, this residence seeks to harmonize with its surrounding landscape with a minimal environmental impact while accommodating the diverse needs of four generations of occupants. Integration with the site is achieved by setting the building low and shifting floor and roof planes so that it becomes embed into the landscape. As the elevations shifts between the deck, loft, upper patio, green roof and floor levels; intimate moments are created as each space unfolds distinctly into the external environment. The elevation of the reflecting pool is also set specifically to create an effect of seamlessly extending Georgian Bay to the house, blurring the lines between building and landscape Material choice is also significant in addressing the issues of site responsiveness, sustainability and for comfort. The rich wood in of the interior spaces brings warmth to the spaces in winter while the rich red cedar and cherry woods intensify the vibrant green of the surrounding forest in summer. The exterior Ipe naturally blends with the surrounding rock soil, and becomes a geometric topography that extends its contemporary aesthetic into the natural environment.

1. living
2. dining
3. kitchen
4. recreation
5. study
6. master bedroom
7. service
8. courtyard
9. garage

first floor plan
一层平面图

Flanked by hard edges on the east and west, the building maintains privacy from neighbours while emphasizing the views to the inner courtyard. The organization of the building is broken down into zones; the main floor is divided between the public / living area and service while upstairs there are the two wings – one for the master suite and the other for the guest wing to accommodate extended family. These spaces are unified by a central courtyard, which acts as the hub of the house, which gives access to all main spaces on the ground floor.

The building has plenty of sunlight. The continuous clerestory windows in the main pavilion offer daylight as well as a 360° view of the adjacent external environment while its upward sloping roof utilizes stack effect ventilation.

The building implements a ground-loop geothermal system as the primary heating system. Comprehensive sustainable technologies and building practice sets this project apart from others in the Blue Mountain region while the unique building form and functional program promise to provide a wonderful family retreat as well as a year round dwelling for its owners.

坐落于Georgian湾南岸的Christie海滨住宅旨在完美地融入其周围的自然环境，并且不产生任何消极的生态破坏，同时，满足其周围居民各种不同的需求。为此，该建筑采用低矮式造型，以及能够自由活动的地板和绿色的屋顶。甲板、阁楼、天井、屋顶，以及各个楼层的此起彼伏，好像其中的每一个空间都连接着户外的自然景观，洋溢着清新、活泼的乡土气息。Christie海滨住宅与其在水池中的影子相映成趣，模糊了各个建筑原本清晰的线条，顿时，好像一片片"秋水共长天一色"。室内纯天然的木材在冬日里带来温暖，而红雪松和樱桃木在夏日里与户外森林的绿意盎然形成鲜明的色彩对比。其外表层的液相外延与周围掩饰土壤融为一体，塑造了具有当代美学价值的地质景观。

该建筑的东西边缘轮廓鲜明，保持良好私密性的同时，强调了由户外到室内的视觉连贯感。该建筑拥有简明的分区：大厅中依次排列着公共空间、居住空间和服务区，楼上则是主人套房和客人房。作为视觉焦点的庭院，恰好位于整栋建筑的中央，将所有空间有机地连为一体。

该建筑拥有充足的自然采光，其连续高大的落地窗从360度各个方向将窗外的自然风光引入室内，其陡坡状的屋顶保证了良好的通风。

在能源供应方面，该建筑配备了地热系统，作为供热的首要选择。可持续性先进技术的综合利用，使Christie海滨住宅在该地区中脱颖而出。其独特的建筑外观和功能齐全的基础设施为在此度假的家庭提供了一个温馨、舒适的休闲天堂。

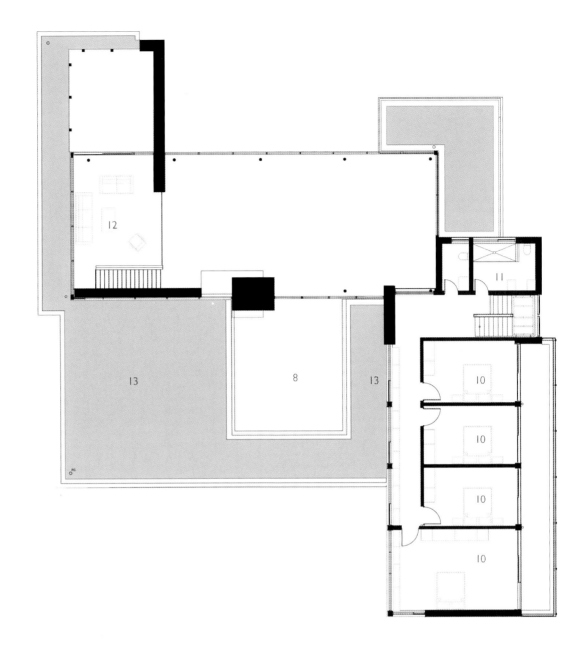

1. bedroom
2. WC
3. loft
4. green roof

second floor plan
二层平面图

Architect: Zerafa Studio Llc.

Location: Niagara Falls, Ontario, Canada

Site Area: 7,400 m²

Gross Floor Area: 500 m²

Photographer: Tom Arban

NIAGARA RIVERHOUSE

The house is comprised of three distinct horizontal volumes, each with a specific material quality. The building's north south massing is defined by two overlaid rectangular shells within which the glass, cedar and granite clad volumes for the interior living spaces are placed and a series of remaining voids create covered exterior spaces. The shell exteriors are clad in silver metal panel and are mostly opaque to provide privacy from adjacent properties to the north and south. The ground floor is comprised of two primary program groups separated by an east west glazed circulation space that bisects the house and extends the river views through to the rear garden.

The service and ancillary spaces, garage, storage, guest suite and access to the basement level are contained within a single-storey bar that runs east to west to minimize the obstruction of views. The primary living spaces are distributed in a linear bar across the width of the site to maximize exposure to river views. The home office, kitchen and dining room and double-height living room extend the full width of the north-south bar to mediate the front and rear gardens and establish a strong visual connection to the outdoors.

Niagara河边别墅由三个独立的水平空间组成，每个空间都由特色材料建成。该别墅由南向北由两个方形覆盖物建成，室内空间的建筑材料为玻璃、杉木和花岗岩；一系列建筑物之间的剩余空隙形成了一些带有顶棚的室外空间。该别墅的外部框架由银色的金属板覆盖而成，这些不透明的金属板为该别墅提供了私密空间。底层由两个主要的建筑系统组成，这两个系统被一个东西方向的流通空间分隔开来。流通空间把该别墅分成了两部分，视线穿过其中，能够从后花园直观河流的景色。

服务及附属空间、车库、储物室、客房和地下一层的入口均建在一个单独楼层里，此处的栅栏呈东西方向延伸，减少了视线的遮挡。主要生活空间的分布横跨了整个场地，站在这里，放眼望去，就能将整条缓缓流淌的小河尽收眼底。家庭办公室、厨房、餐厅和双层高的起居室，从南栅栏到北栅栏，形成前后花园的天然衔接和自然过渡。

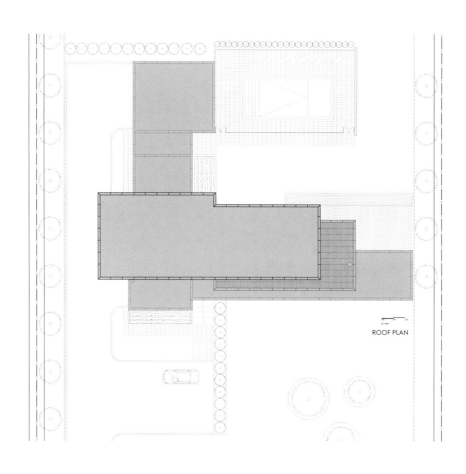

roof plan
屋顶平面图

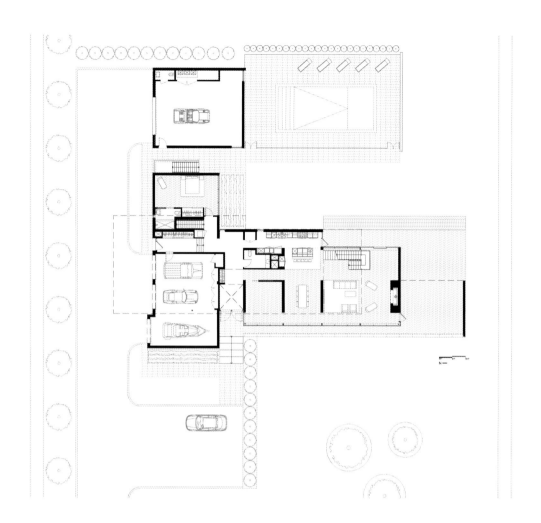

first floor plan
一层平面图

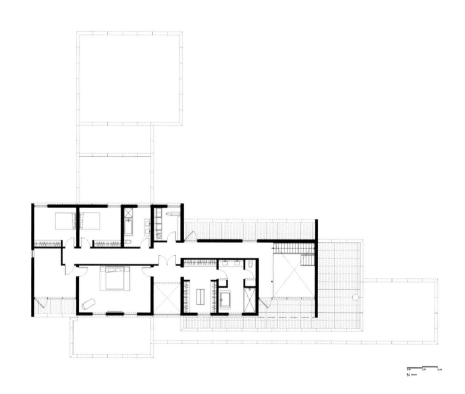

second floor plan
二层平面图

The more private spaces including an expansive master room, two additional bedrooms with en-suite bath and laundry facilities are distributed in a parallel bar on the second floor accessible by a dramatic sculptural stair. The master room extends the length of the river view façade, bridging across the circulation space below and extending out to a large covered terrace. The double-height living room features a custom-designed stair, engineered and fabricated locally of steel-reinforced maple, which is full of old and elegant fragrance of musuem.

宽阔的主卧具有良好的私密性，两个带独立卫生间的卧室和洗衣设备分布在可以由雕塑楼梯进入的二楼平行栅栏里。主卧一直延伸至河流的尽头，并且横跨下面的流通空间，蜿蜒曲折地连接着一个带有顶棚的平台上。在双层高的起居室中，有一个别出心裁的楼梯，它由取材于当地的杉木建造而成，彰显着"博物馆式"的古老和典雅。

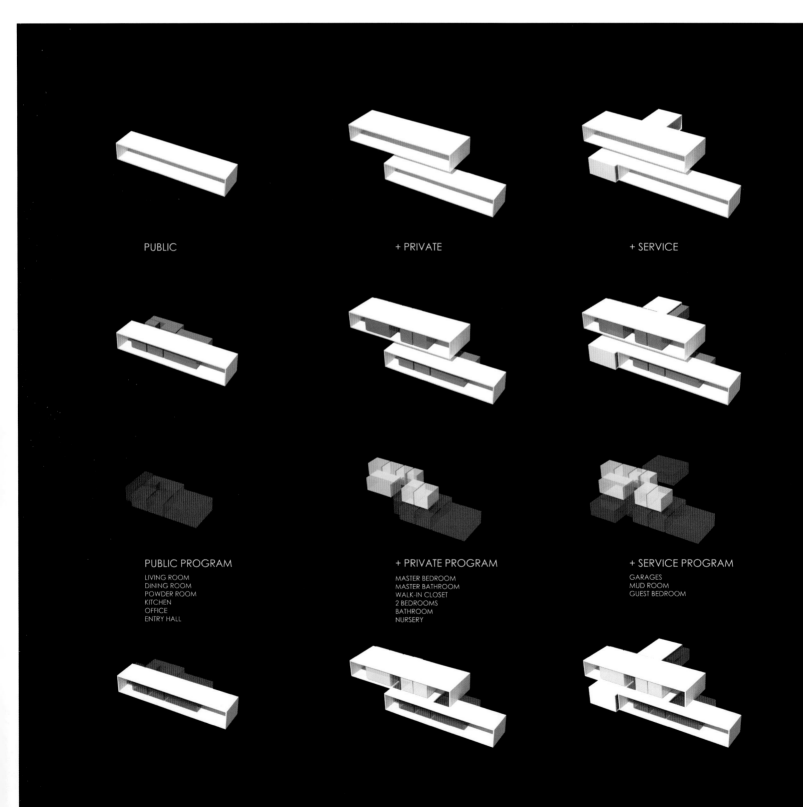

PUBLIC　　　　　　　　　　+ PRIVATE　　　　　　　　　　+ SERVICE

PUBLIC PROGRAM
LIVING ROOM
DINING ROOM
POWDER ROOM
KITCHEN
OFFICE
ENTRY HALL

+ PRIVATE PROGRAM
MASTER BEDROOM
MASTER BATHROOM
WALK-IN CLOSET
2 BEDROOMS
BATHROOM
NURSERY

+ SERVICE PROGRAM
GARAGES
MUD ROOM
GUEST BEDROOM

Architect: YH2_Yiacouvakis Hamelin Architects
Builder: Martin Lachance
Location: Saint-Hyppolite, Quebec, Canada
Project Team: Benoit Boivin, Marie-Claude Hamelin, Loukas Yiacouvakis

GEOMETRY IN BLACK

In the Laurentians, a dense forest on a slight hill, down-turns into the expansion of a small river. Through the trees, the body of a black building is divided into three blocks linked by glass passageways. Three blocks of a home, mid-level from each other, are all in direct contact with the earth. Three blocks of proper identity, offering intimacy between each and open to nature: An entry block, open on two levels and includes the adolescents quarters and family room. A daytime block, central space, friendly, opens onto the terrace. A private block, owners suite, isolated from the rest of the home.

在Laurentians地区的一片茂密的森林中，屹立着一座小山，其山脚连接着一条小河。穿过森林，一座黑色的建筑映入眼帘。

该建筑分为三栋高楼，由玻璃通道连接而成。它们风格统一，均成中等高度，与地表直接相连，代表着三种不同的建筑理念——各具特色，彼此亲密，亲近自然。入口大楼包括两层，分别为青少年空间和成人空间。白天大楼是整栋建筑的中心地带，令人感到亲切、友好，其中还有一个开放的露台。私人领域是与其他空间分离的主人套房，宁静且安逸。

On the north side of the house, a large section of bent corten steel with oblique lines connects the fragmented blocks together while defining a series of outdoor settings, always against the light. Partly-based in the geometric basis of the concept which dictates each project component, a sort of unwritten contract exists between architect and client. The modification of each element having a direct influence on the others, the work of enlarging or reducing a block in height and plan cannot be achieved without enlarging or reducing other areas of the house.

This geometry which is both fragmented and linear makes the project a strong spatial experience, allowing direct and variable contact with the landscape. It is this angular part, made of dark and raw materials that unites the house to nature, like a rock that emerges from the ground, or a forgotten shipwreck at the heart of the forest.

在整座建筑的北部,柯尔顿钢板的线条呈倾斜状,将三栋原本割裂的大楼连接起来,从外面看来,在自然光的照射下,俨然一个整体。它们表面上各处一隅,但实际上,彼此又有着千丝万缕的联系。它们在外部景观上的相互作用,使任何一个对于细节所做的改动,都必须从全局着眼。

这种既割裂又呈线性的结构设计,赋予该建筑以强烈的空间感,同时与外界形成直接且变幻多样的自然联系。该建筑仿佛从地平线上突起的一块岩石,又好像一座处于森林中心的、被遗忘的船骸。

Architect: Blouin Tardif Architecture-Environnement
Landscape Architect: Projet Paysage
Structural Engineer: CLA Experts Conseils
Project Manager: Alexandre Blouin
Project Team: Isabelle Beauchamp, Sophie Martel, Jonathan Trottier, Alexandre Blouin
Location: Rawdon, Quebec, Canada
Photographer: Steve Montpetit

LA SOURCE – MASSAGE THERAPY PAVILION

This new pavilion is part of a resort grouping built on a mountainside in the Lanaudière region. The project as a whole is integrated into nature through its siting, the use of natural materials, and the framing of the surrounding landscape through generous openings in the contemporary architectural structure.

The construction of this building with an area of 800 m² has three levels. The facilities now include thirteen massage spaces and two new lounges. In addition, the new pavilion integrates a reception area for customers and administration offices. This combination of functions, requiring both a contemplative ambience and a workplace, represented not only one of the challenges for the project but also the most significant concern to the designer.

The project is designed as a procession featuring the surrounding natural environment. Visitors arrive at the pavilion via a long wooden walkway overlooking the forest. This outdoor passageway leads to the entrance while establishing a dialogue with the adjacent escarpment. Situated in the centre of the building and framed by rough-concrete walls, the main stairway provides access to all levels.

The building's cladding is made of torrefacted poplar; torrefaction is a procedure that makes the material (a local wood species) very strong and durable.

La Source按摩疗养院坐落于Lanaudiere地区的一座小山上，其周围是一片度假村。该建筑整体上亲近自然，使用取自于当地的建筑材料，利用当代建筑结构的开放空间，搭建其周边的自然景观。

La Source按摩疗养院占地800平方米，是一栋3层楼的建筑，包括13个按摩间，两个新建的大厅，一个顾客接待区以及众多的行政办公室。打造高度功能性兼具自由思索型的休闲以及办公空间，是该建筑面临的最大挑战，也是建筑师的首要目标。

该建筑以"与自然和谐共生"为特色，顾客可以通过一条能够鸟瞰森林的木质走道进入其中。该走道将疗养院的入口和其周围的陡坡连为一体。通往每一层的主楼梯恰好位于疗养院的中心，被原生混凝土墙面所包围。

该建筑的外包面由美国鹅掌楸木装饰而成。经过"烘焙熔蜡"程序处理的鹅掌楸木坚固耐用，具有长久的生命力。

The green L-shaped concrete roof orients the pavilion toward the mountain and the forest. The roof, like a big umbrella, protects the building envelopes, the fenestration, outdoor walkways, and the patio. It has been polished and left visible to minimize finishes and maximize the thermal mass effect. The radiant-floor heating system is deployed in all spaces, providing comfortable, efficient warmth to users.

This project is intended first and foremost to be a space of relaxation and contemplation in the midst of nature. It integrates the values proposed during the design process: the comfort of the occupants through the use of warm, natural materials with simple forms, contact with the surrounding landscape, and a structure in harmony with nature.

绿色混凝土"L"形屋顶确定了整个疗养院的大致方向——面朝山区和森林，可以呼吸新鲜的空气。它好像一把"大伞"，有力地保护着疗养院的外包面、天窗、外部走道以及天井。除了被抛光，该屋顶未做任何装潢或粉饰，保留"原生态"的同时，实现"蓄能效应最大化"。每个房间均配备了辐射式地板加热系统，营造了舒适、温馨的空间氛围。

该建筑充分利用取材于当地原生态建筑材料，顺应周边地形、地理条件，以人为本，将最大限度地为顾客提供惬意且充满情调的疗养空间为根本宗旨。La Source按摩疗养院沐浴在自然之中，是放松身心的疗养胜地，是自由冥想的梦想田园。

Architect: Pierre Cabana
Location: Brompton, Quebec, Canada
Photographer: Richard Poissant et Pierre Léveillé

SANCTUAIRE MONT CATHÉDRALE

The house uses art as much as architecture to blend into its surroundings. Inspired by a tree, the house's first floor is covered in wood to represent a tree trunk. The much larger second floor is covered in copper tiles (leaves), which will, in time, develop a green patina, helping the house harmonize with the surrounding vegetation. An immense glass wall faces southeast on the front of the house and opens onto a 70 m² terrace looking out onto a panoramic view of the lake and neighboring mountains. The terrace is 54 m above lake level, creating a sense of weightlessness and direct contact with the natural elements.

该建筑堪称一件与其周边环境完美融合的"艺术作品"。它的一层处在一片森林之中,二层的地面由加拿大铜瓦铺设而成,在此之上,搭建一个绿色的铜盘,作为该建筑与户外景观融为一体的标志。大面积的明亮的玻璃墙面朝西南,将人们的视线引向远处的湖泊和山脉。70平方米的平台高出湖面54米,站在这里,就能将周围的自然景观尽收眼底,美不胜收。

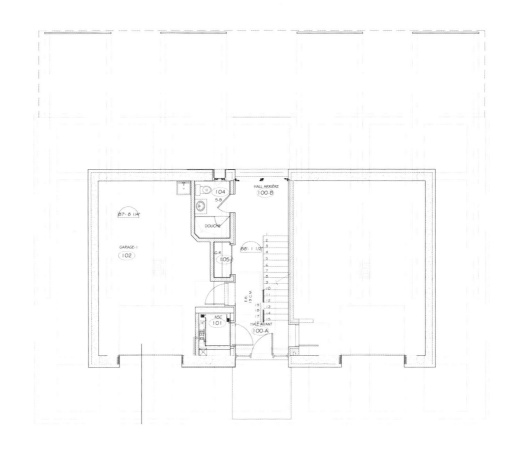

first floor plan
一层平面图

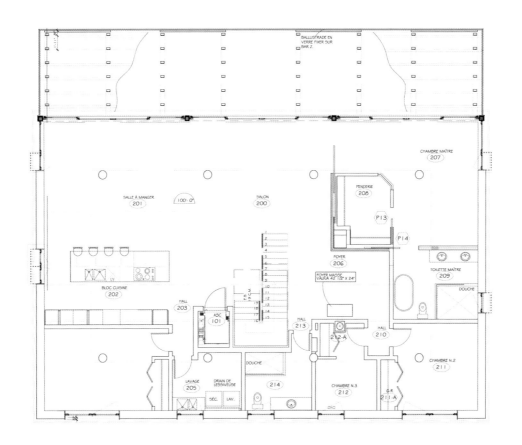

second floor plan
二层平面图

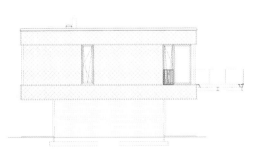
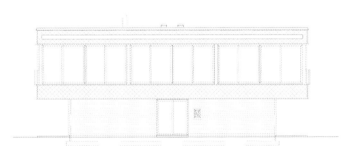

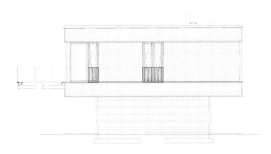
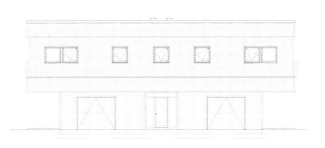

Inside, the house is friendly and warm thanks to the judicious use of natural and manufactured materials. The marriage of wood, marble, quartz, steel, glass, and porcelain creates a harmonious connection with the natural environment.

Insulated with great care, the house has excellent energy efficiency. The heating system uses a combination of geothermal energy. In addition, the glass wall increases the energy efficiency of the house. In winter, sunlight comes in and reduces heating costs; in summer, electric, automatic awnings keep the interior cool.

室内空间充分利用当地原生态的建筑材料，例如，石英、钢铁、玻璃、陶瓷等，洋溢着浓郁的自然气息。

该建筑在节能方面十分高效，地热被充分利用来为空间供暖。另外，得益于大面积的玻璃窗，自然光线能够充分地照射进来，降低了冬日里的供热成本；夏天，屋顶上的凉篷自动开启，保持室内的凉爽和舒适。

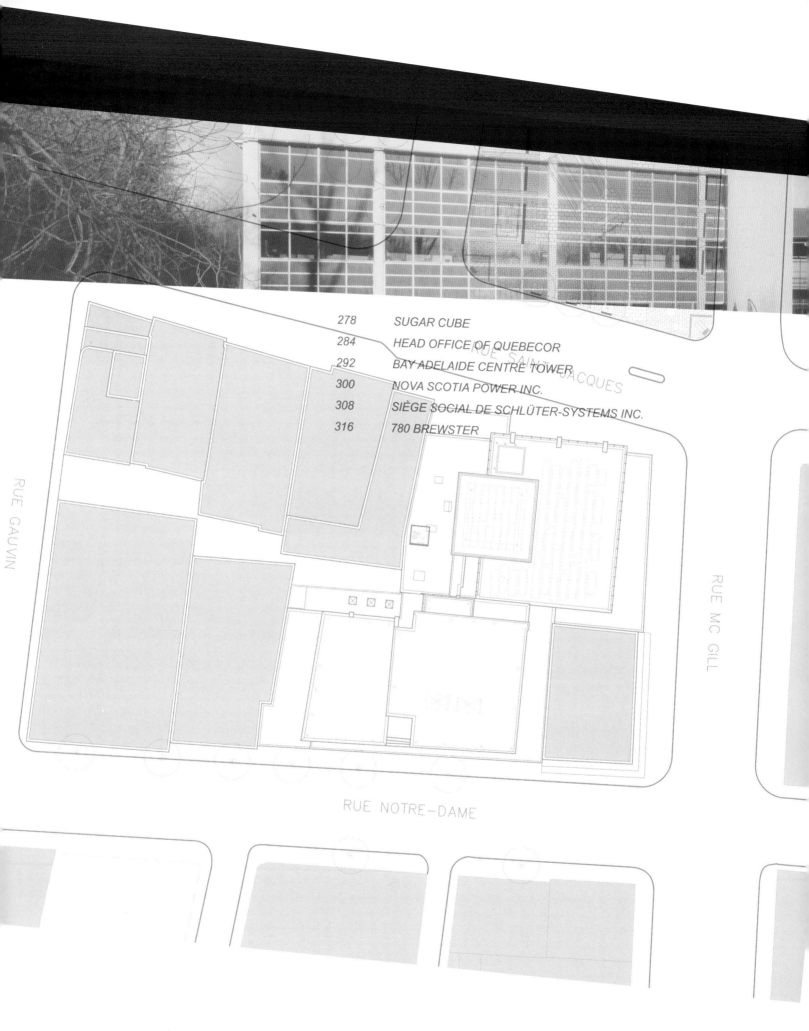

278	SUGAR CUBE
284	HEAD OFFICE OF QUEBECOR
292	BAY ADELAIDE CENTRE TOWER
300	NOVA SCOTIA POWER INC.
308	SIÈGE SOCIAL DE SCHLÜTER-SYSTEMS INC.
316	780 BREWSTER

Architect: KPMB

SUGAR CUBE

This mixed-use development project introduces a new strategy for making contemporary architecture within Denver's historic Lower Downtown Heritage District (LoDo).

The client's specific functional and performance goals included building for longevity as a form of sustainability, achieving cost savings with energy efficient systems and enduring design, and creating an active street-related base to contribute to the revitalization of the district.

坐落于丹佛LoDo历史文化中心的Sugar Cube建筑具有多种用途，堪称运用当代新式建筑技术的经典之作。

客户对于建筑质量的要求非常高，希望该建筑能够具有可持续性，并且经得起时间的考验，实现降低成本、节约资源和独具创造性的"一箭三雕"。

The project is located on the 16th Street Mall, a major public pedestrian thoroughfare that runs through the city of Denver. The building features a central ten-storey volume in manganese-colored brick, wrapped around by buff brick. The top of the parapet of Sugar Cube is set at a height that aligns with the underside of the upper cornice of the Sugar Building. The roof provides generous and bright outdoor terraces to show the beauty of the whole city.

The darkness of the cube's brick and the way it interacts with Denver's strong light creates a dramatic contrast with the lighter masonry volumes, and inserts an iconic, modernist form on the Denver skyline. The scheme explores alternatives to conventional balconies associated with residential developments to create drawer-like projections from the building, and make an unusual interplay of form against Denver's brilliant blue skies and distant views of the Rocky Mountains.

该建筑是丹佛LoDo历史文化中心内的第16号大厦，人们可以通过一条贯穿该城市的公共人行通道进入其中。建筑物内墙由锰色瓦砖铺设而成，外墙则由软牛皮瓦砖包裹而成。该建筑顶层的护栏与下方建筑的边缘精确地保持一致，露天阳台宽敞且明亮。站在这里，低头俯瞰，能够将城市美景尽收眼底。

由锰色瓦砖和软牛皮瓦砖营造出来的深色光线与丹佛整个城市明亮的自然光线形成鲜明的对比，仿佛形成一条标志性的丹佛地平线。建筑师摒弃了传统露台的建造方法，而另辟蹊径，将露台打造成抽屉的形状，巧妙地位于建筑一隅，就好像建筑头戴一顶美丽的大帽子，在蔚蓝的丹佛天空中，在遥远的落基山脉周围，又形成一道独特而亮丽的风景线，清新自然、夺人眼目。

Architect: Cardinal Hardy / Le Groupe Arcop Architectes en Consortium
Location: Montreal, Quebec, Canada

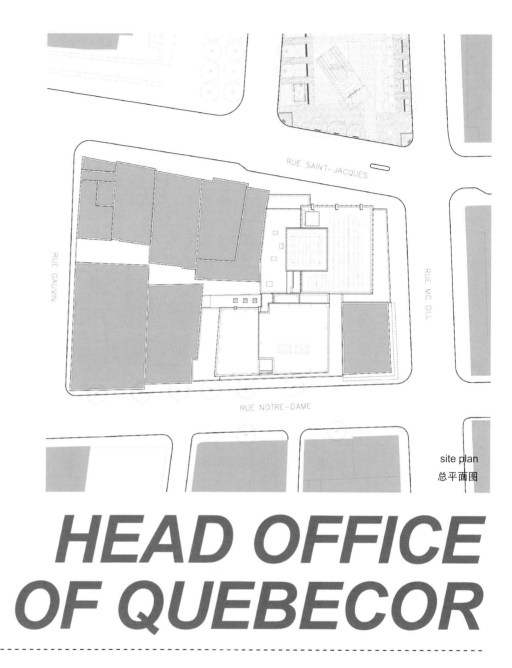

site plan
总平面图

HEAD OFFICE OF QUEBECOR

At the crossroads of Old Montreal and the Quartier international de Montréal, Quebecor's new head office is set in an architectural context punctuated by different eras. limestone-clad base blends with the horizontality of Rue Notre-Dame, topped by a glass-wall tower that dialogues with the adjacent towers, and the twenty-one-storey building is harmoniously wedded with a dense and relatively complex urban fabric. Seen from Victoria Square, the tower's sharp, atypical silhouette matches the urban landscape.

The present project involves an expansion of 170,000 m², with the south side abutting the existing 13-storey building by an internal passageway.

As a natural continuity with the existing building, three conditions seemed essential to the architects: How to connect the two buildings to the urban context and environment in every sense? How to adopt refined approaches to urban design and project management? How to achieve the client's needs of an economical, simple project in a context of performance and long-term value, based on the high-quality materials?

Quebecor新总部位于蒙特利尔新旧城乡的交汇处。坚固的石灰石地基平行于圣母院街的地平面。从维多利亚广场望去，这栋21层的"玻璃墙"大厦，其锋利且不规则的外形在其周围的建筑群中"脱颖而出"，成为一道神奇的都市建筑景观。

Quebecor新总部以原有的13层大厦为基础，在其周边17万平方米的土地上新建一座大厦，两栋大厦在南面由一条内部的人行通道连接起来。

建筑师面临三个方面的挑战：如何将这两栋建筑完美地融入其周边的自然环境？如何在设计和管理方面相比以前，有所创新？如何满足客户提出的"建筑材料经济、简约、持久、耐用"这一要求，打造高质量的建筑？

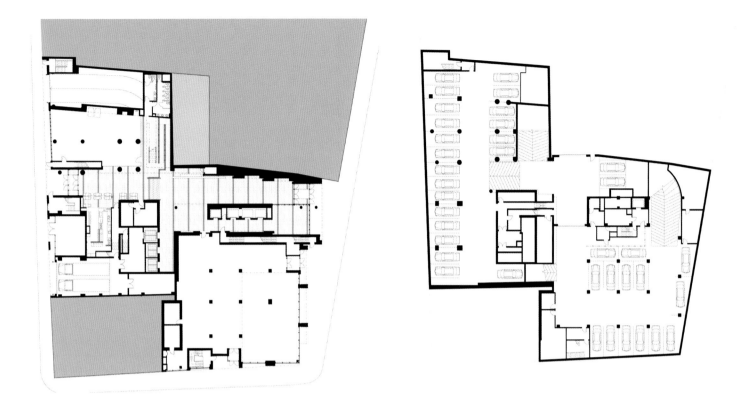

Therefore, the idea of making full use of the structure of the existing building by grafting a second building in continuity with it was the beginning of a long series of principles that led the architects to think about their project around the concepts of recycling and reinterpretation of the existing architectural elements. This dynamic of linking the two projects is very perceptible in the color and texture of the façades, in which volumes of glass and masonry stretch vertically, meeting or extending to the top. Limestone, black granite, glass and aluminum are used to create a neutral-colored building alliance.

The architect has introduced a number of green roofs with sharp forms, enclosing a small garden. They effectively reduce the heat-island effect retain rainwater, improves the lifespan of the waterproofing membranes, and encourages biodiversity with diverse flowers and trees. Although the idea was set aside due to budgetary constraints, it remained with the owner, who finally gave it the go-ahead as the project was being completed.

因此，对原有建筑结构及材料的再利用和重新解读就成为该设计的重中之重。新旧建筑的外立面在材料和色彩搭配方面都极为相似，玻璃砌石呈垂直堆砌状，直达甚至越过了顶层。石灰石、黑色花岗岩、玻璃、铝合金的色彩运用恰到好处。远远望去，新旧大楼俨然就是一个整体。

新总部的顶层建有许多尖状的绿色屋顶，好像一把把直插天空的"尖刀"，令整栋建筑"高耸入云"，而众多的"尖刀"又围成一个小花园。这样的设计有利于散发热量，抑制"热岛效应"；同时在雨季也方便将倾斜的大量雨水完整地收集起来，回收利用，更重要的是，它有效防止了雨水对建筑材料的腐蚀，延长了防水建筑薄膜的使用寿命；其中种植的花花草草，则促进了生物多样化，呈现一派"绿意盎然"的景色。虽然受到了预算资金的限制，但客户的鼎力支持，为建筑师完成这些"锋利的尖刀"扫除了障碍。

On the ground floor, the creation of a new entrance hall takes advantage of the relocation of the old parking-access ramp. Thus, a new two-storey-high public passageway crosses the two buildings from one side to the other, connecting Rue Notre-Dame to Victoria Square. In its centre, a reception area formed of a long furniture wall of black granite faced with glass directs the public toward the vertical circulation and services spaces, creating a sense of horizontality, transparency and airiness.

All of the interior spaces in the project are designed, from the beginning, as obstacle-free and universally accessible environments, with a lot of service infrastructure convenient for the disabled. Benefiting from the greatest autonomy possible, complete freedom and safety will be enjoyed.

Quebecor's new head office respects a defined urban architecture style, innovative, functional and tasteful.

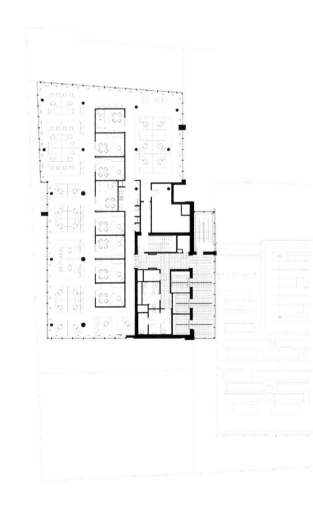

在一楼，建筑师将原先的停车场斜坡拆除，在此新建一个宽敞的入口大厅，其中包括一条两层高的贯穿新旧大楼的人行通道，另一端则通往圣母院街和维多利亚广场。接待区处于大厅的中央，被黑色花岗岩和玻璃搭配而成的墙面所包围，并且毗邻流通区和服务区，增强了大厅的水平感、透明度和通透性。所有楼层的室内空间均采用"无障碍"设计，进出方便，相互贯通，并配备了许多方便残疾人士的人性化服务设施。在这样的室内空间中工作、休息，令人倍感自由与舒适。

Quebecor新总部代表了一种都市建筑风范，将创新、实用和高品位融为一体。

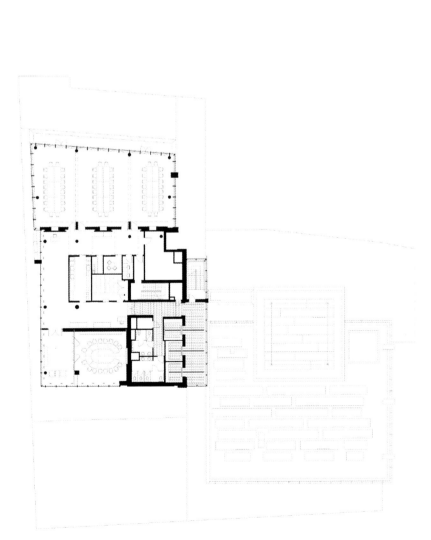

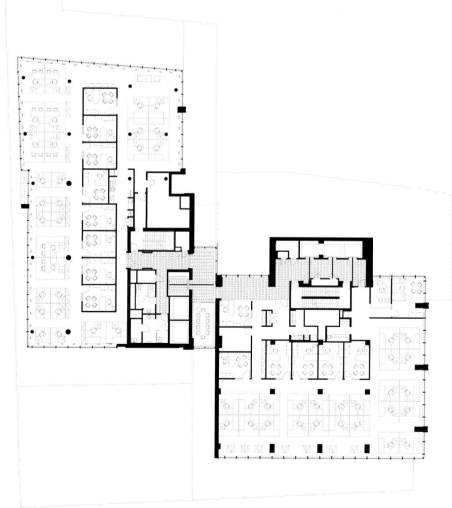

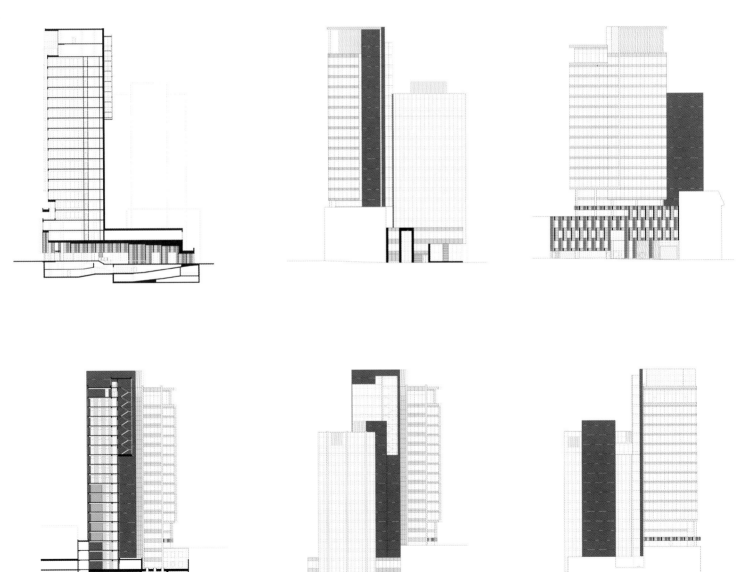

Architect: WZMH Architects
Landscape Architect: Dillon Consulting
Interior Architect of Base Building: WZMH Architects
Structural Engineer: Halcrow Yolles
Location: Toronto, Ontario, Canada
Area: 111, 000 m²
Photographer: Tom Arban

BAY ADELAIDE CENTRE TOWER

  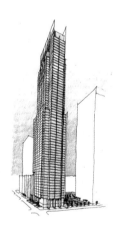 

The 51-storey Bay Adelaide Centre Tower, completed in January 2010, is located on the western edge of a development site that occupies two city blocks, in the financial core of the City of Toronto. The project contains over 111,000 m² of rentable class-AAA office space and includes over 3,700 m² of below-grade retail space linked to the downtown's path system.

At the corner of Bay Adelaide Street, the highly transparent main building lobby, with walls clad in classic Statuario marble and Makore wood, engages passersby. At night, the illuminated lobby becomes a "beacon" at the corner of Bay and Adelaide Streets.

2010年1月建成的Bay Adelaide办公中心大厦位于多伦多西部边缘，共有51层，占据多伦多参政中心两个街区的面积。其拥有11万平方米的顶级办公空间和3 700平方米、经济实用的零售空间，并且与周围道路连为一体，简直就是一个基础设施齐全的小都会。

该大厦位于Bay Adelaide街区一隅。其透明的大厅以及由大理石和木头镶嵌而成的墙面，吸引路人在此驻足。夜晚，被点亮的大厅成为该街区明亮的灯塔。

The project is a modernist building inspired by and paying homage to, the distinctive character of the architecture of Toronto's Financial Core. More transparent than any other in the downtown, the tower is a pristine glass prism clad in clear vision glass and spandrel panels with ceramic frit. The glazing is supported by four sided structural silicone within a channel surround creating a sense of lightness and delicacy for the building skin.

At the top of the tower, the curtain wall extends beyond the roof to become a series of "sails" that create a distinctive silhouette on the city skyline.

　　该大厦洋溢着浓郁的现代气息，其风格沿袭多伦多财政中心大楼的"庞大而不臃肿"的模式。该大厦在其周围的建筑中是最最"透明"的。它由玻璃黏土包装而成，呈"拱肩"的形状。四个基柱均由硅树脂支撑，仿佛坚定、有力的四只大脚，稳稳地扎进土壤中。矗立一方，俨然一个晶莹剔透的三棱镜积木，将这座城市的闪光点毫不吝啬地集于一身，小巧精致，令人爱不释手。

　　该大厦顶层的防护墙伸出了许多"触角"，好像扬帆起航的帆船，形成一道独特的空中风景线。

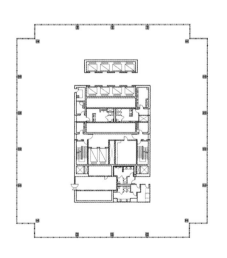

typical plan
标准层平面图

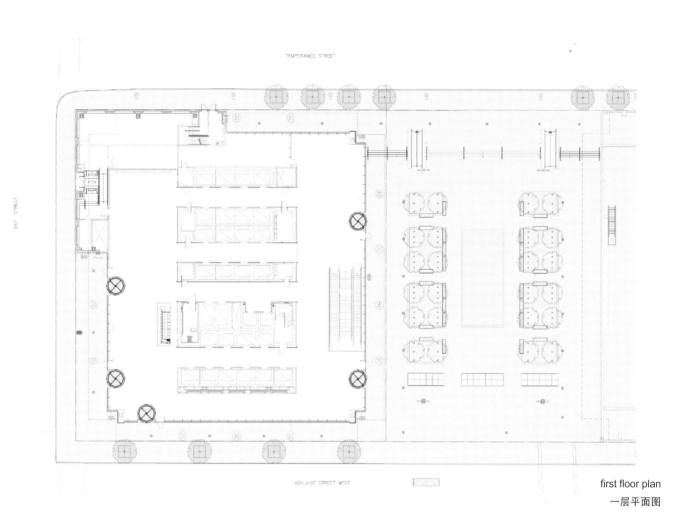

first floor plan
一层平面图

The floors and the plaza are clad in a "carpet" of Brazilian Ipanema granite expressing a modernist sensibility for spatial continuity from inside to out.

Respecting the formality of the tower the plaza's design is simple, comprised of a central lawn framed by planting beds of natural grasses with two double rows of Ginko trees and comfortable seating benches.

The project provides flexible and cost-effective design model, emphasizes energy conservation, and follows sustainable principle. It successfully creates a secure, efficient and comfortable working environment.

购物广场的地面及其室内的地面均由巴西伊帕内玛花岗岩铺设而成，使室内、室外自然过渡、浑然一体，彰显现代与理智。

外部设计依然遵循整体风格的简约、淡雅，草坪上铺满了本地生的嫩草，种植着双排Ginko树。舒适的长椅供人们休息、乘凉。

该大厦采用灵活、低成本的设计模式，注重节约能源，遵循可持续发展的环保原则，从而打造了安全、高效、舒适的办公环境。

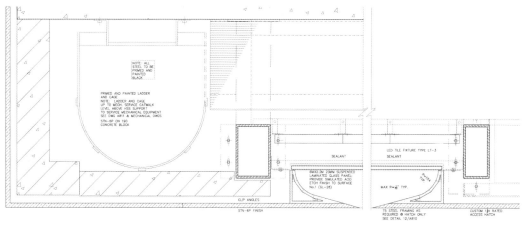

PUBLIC ART— PLAN DETAIL

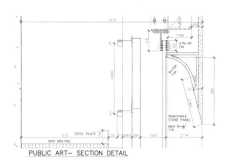

PUBLIC ART— SECTION DETAIL

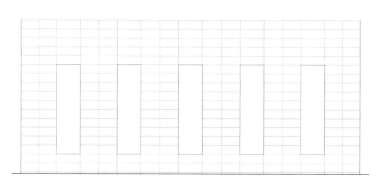

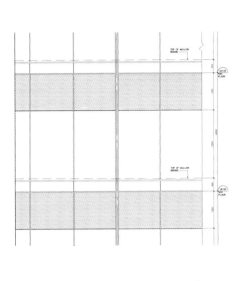

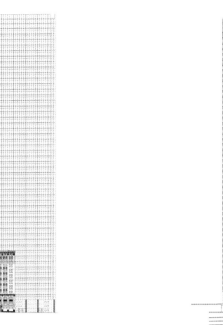

curtain wall detail
幕墙详图

west elevation
西立面图

north-south section
南北剖面图

Architect: WZMH Architects

Landscape Architect: Gordon Ratcliffe Landscape Architects

Structural Engineer: BMR Structural Engineering

Location: Halifax, Nova Scotia, Canada

Area: 17,930 m²

Photographer: RPM Productions

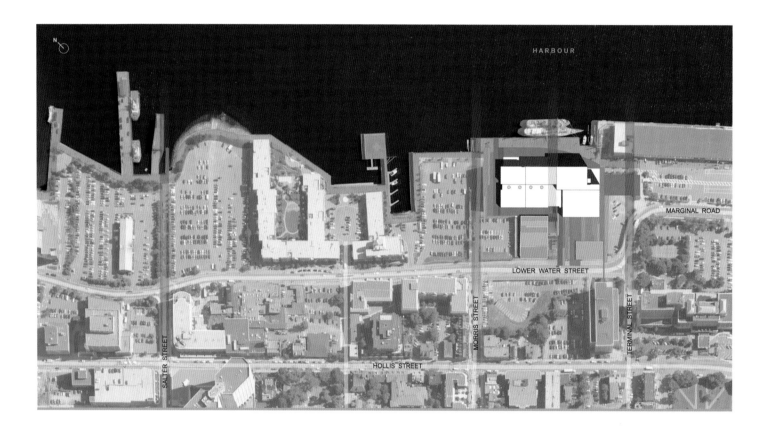

NOVA SCOTIA POWER INC.

The project is located on a 5-acre site at the southern end of the Halifax downtown waterfront with access from Lower Water Street. The site steps down approximately 7.5 meters from Lower Water Street to the harbor, east of the site. To the south, there has been significant redevelopment of some of existing harbor buildings. To the north and west, vacant lots exist that will be subject to future development.

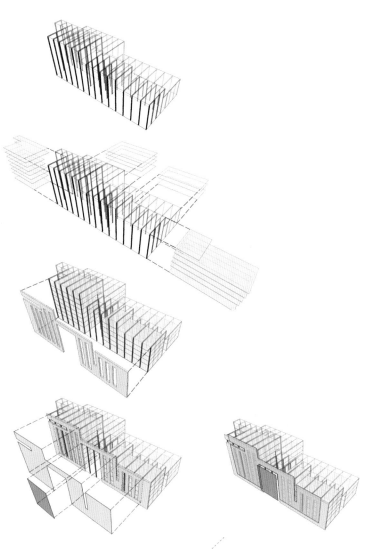

The project involves the retention and adaptive reuse of the former generating plant to become the headquarters for the provincial electrical utility, Nova Scotia Power Inc. (NSPI). The facility will house over 500 staff in approximately 18,000 gross m^2 and provide onsite parking for 150 cars.

Early visits to the site were inspirational: soaring interior spaces with an exposed latticework of steel framework were reminiscent of the imagery of Russian Deconstructivism design. Investigation of archival photos of the site uncovered the image of a line of 4 chimney stacks, becoming unique skylines.

As the provincial power authority, NSPI wishes to demonstrate environmental responsibility and show leadership in energy conservation. The unique adaptive reuse of the building will be a visible statement of the corporation's

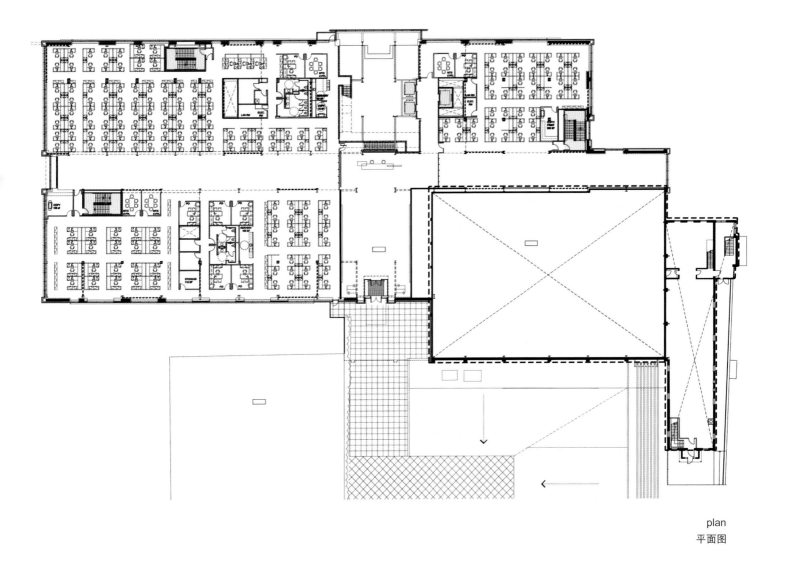

plan
平面图

303

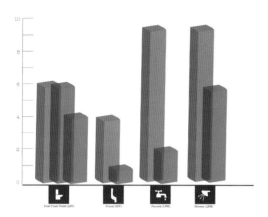
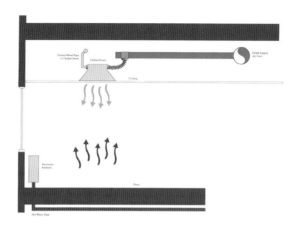

commitment to sustainability. Sea water cooling (and heating) is being provided utilizing existing piping from the Halifax harbor originally used to cool power generating turbines. The building will represent the first major use of "chilled beam" technology in Canada. Located within the ceiling space, the system, more efficient than conventional systems, utilizes (low energy sea) water rather than air to transport cooling thereby lowering energy consumption.

Additional energy saving strategies include the provision of heat recovery on HVAC, daylight and occupancy sensors for lighting and supplemental heating for both the building and hot water with the use of solar thermal panels. NSPI deserves to be the example model of the energy-saving industry.

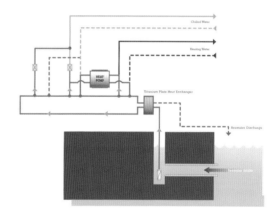

NSPI发电厂位于哈里法克斯市中心海滨南岸的5公顷的土地上，从这里能够直达Lower Water街区，再往前走7.5米，就到达海滨东岸的港口了。它的南面是港口大楼的所在地，从事进出口贸易；北面和西面则是尚待开发的土地。

这次改建，旨在将该发电厂打造成该省电力机构的总部。它将容纳500名员工，办公空间占地1.8万平方米，停车场将能够容纳150辆汽车。

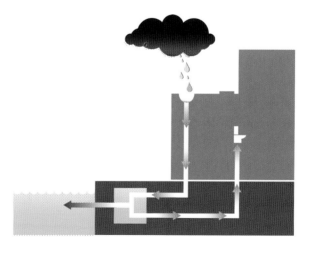

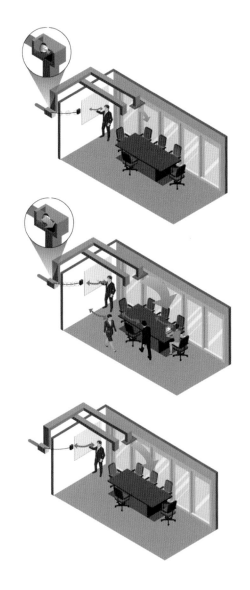

　　方格式的入口，仿照俄罗斯"反构成主义"建筑风格。一进入室内，空间高度急剧上升，视野极为开阔。4根铝合金的大烟囱，各自独立，互不干扰，蜿蜒起伏地延伸至室外，形成一道独特的风景线。

　　室内装潢追求"工业美感"，抛光的混凝土地板，搭配铝合金金属网眼儿的楼梯扶手，亮丽中透着高雅。通往Lower Water街区这一段7.5米的空间，被建造成一个高高的中庭。它不仅成为发电厂和街区的天然衔接，而且站在这里，就可以将远处港口辽阔的海景尽收眼底，大饱眼福。

　　作为该省最具权威性的发电机构，NSPI发电厂始终严格贯彻可持续发展的环保原则，在节约成本的前提下，实现其实用性价值。例如，利用海水冷却和加热技术实现发电效率最大化，并在加拿大境内首次成功地使用"冷却光束"，实现了最小化能源消耗、最大化发电效率的成功突破。其他节能举措包括HVAC热量恢复、利用太阳能热板实现日照和供水的感应开关等等。NSPI发电厂堪称当地环保节能型企业的典范。

Architect: DCYSA Architecture & Design (Anh Lequang, Lucien Haddad, Azad Chichmanian)

Special Collaborator: Lyse M. Tremblay, PA LEED

Photographers: Gleb Gomberg, Alex St-Jean

SIÈGE SOCIAL DE SCHLÜTER-SYSTEMS INC.

Sustainability and design are interwoven in this industrial structure. Technology is celebrated and treated as an integral aesthetic experience. A medium-sized project in surface area, the building explodes with innovation and reads as a rigorous exercise in energy efficiency and contemporary language. The client is Schlüter-Systems, a company that manufactures accessories and membrane products centered on the tiling industry, and thrives on experimentation and perfection.

The building is the Canadian headquarters and serves as an office and training centre of the company's products. The two programs are architecturally distinct and

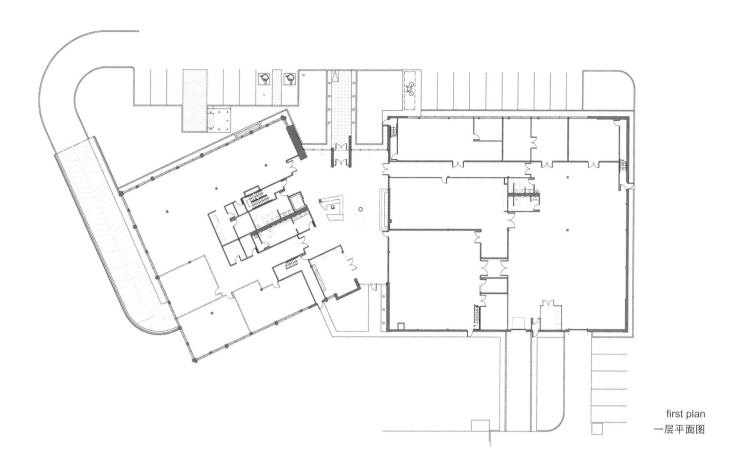

first plan
一层平面图

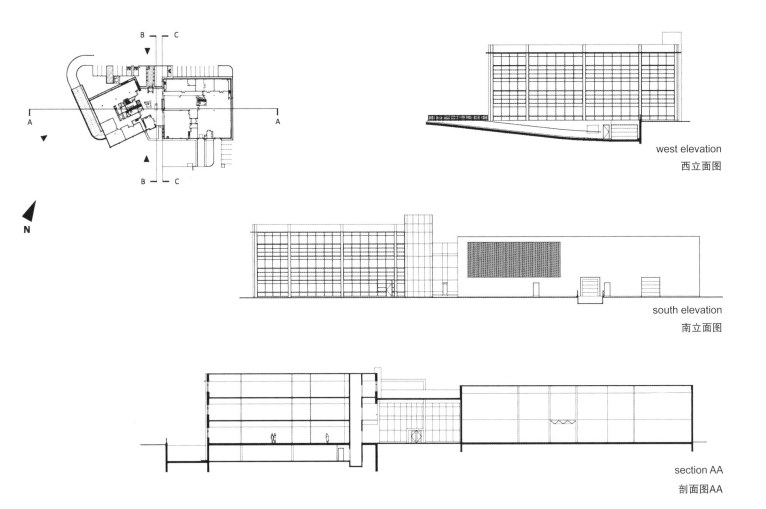

west elevation
西立面图

south elevation
南立面图

section AA
剖面图AA

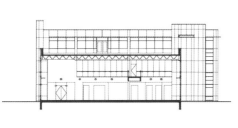

section BB
剖面图BB

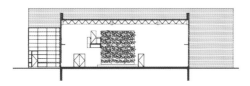

section CC
剖面图CC

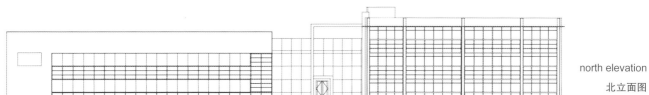

north elevation
北立面图

connected by a delicate glass box, with a bold folded plane in the foreground marking the entrance. The combination of elements – towers, glass, folded plinth – reads as a harmonious exchange of lightness and depth, and points to the expertise of the company. The industrial complex is quiet and distinguished, sitting in a large wooded site in Sainte-Anne-de-Bellevue.

The combination of a dedicated team of specialists and an integrated process leads to a successful LEED-Gold certified building. Moving beyond basic principles of sustainability, the project manages to bring innovation to water management, energy use, and material selection. Ultimately, the building not only celebrates the company it was built for, it promises a long existence of pleasant use and ease of maintenance.

The building stands as a model for the possibilities of a successful business, and points to the future of environmentally intelligent corporate design directions.

"可持续性"是该建筑的特色和主要闪光点。在该建筑应用的先进技术被称颂的同时，其可持续性的节能环保意识和美术设计更令人刮目相看。这是一个中等规模的工程，客户是Schlüter - Systems，专门从事瓦砖工业配件和薄膜的生产。

该建筑被用来作为该公司的加拿大总部和产品开发的训练中心。这两个部门，虽然在建筑风格上各具特色，各成一体，但双方却同时由一个设计精致的玻璃框和呈折叠状的机翼作为入口处的背景，这样，它们又被"鬼斧神工"地联系在一起。塔、玻璃、折叠柱基，三者的结合，自然且和谐，同时也指明了该公司的专业所在。整栋建筑优雅、安逸、风格独具特色，静静地坐落于Sainte-Anne-de-Bellevue的绿树丛林之中。

业务顶尖的专家小组和完整的设计流程为该建筑荣获LEED金奖助一臂之力。该建筑不但超越了可持续发展的基本原则，而且走得更远、更深，它试图在水资源管理、能源使用和材料选择方面做出前所未有的创新之举，最终，它光荣地完成了设计任务，并且，有信心地承诺：在今后很长的时间内，这里定会是一个令员工心旷神怡的工作环境，同时也非常便于保养和维修。该建筑为今后Schlüter-Systems通往成功之路铺平了道路，同时也为环保节能型办公空间设计的未来发展指明了前进的方向。

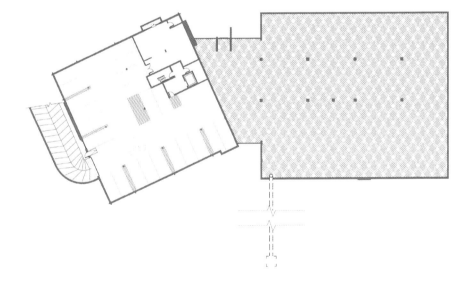

basement plan
基础平面图

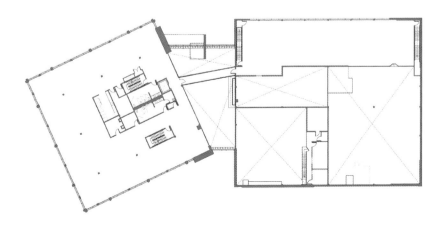

second plan
二层平面图

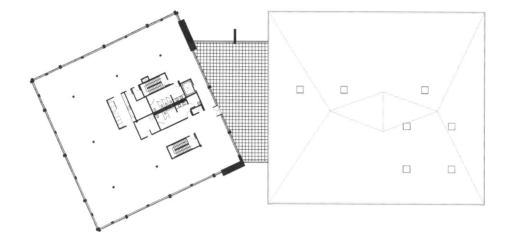

third plan
三层平面图

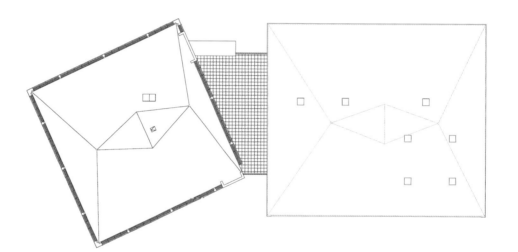

roof plan
屋顶平面图

Architect: LEMAY

Location: Montreal, Quebec, Canada

Area: 12,000 m²

Photographer: Claude-Simon Langlois

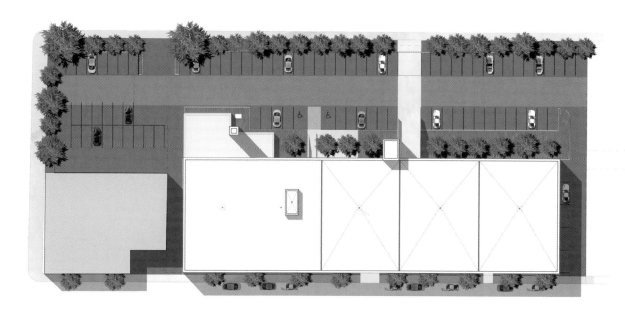

site plan
总平面图

780 BREWSTER

780 Brewster is a five-storey multi-tenant industrial building with a total surface area of 12,000 m². This brick and timber building was transformed into an office building housing, elegant and exquisite.

What deserves to mention is that the structure meets the most rigorous sustainable development standards and is certified LEED Silver.

这是一座5层的多功能办公大楼，占地面积为12 000平方米。该建筑外观在材质上实现了砖和木材的完美结合，素雅且精致。

最值得一提的是，该建筑严格遵循了可持续发展的环保原则，从而荣获了LEED设计银奖。

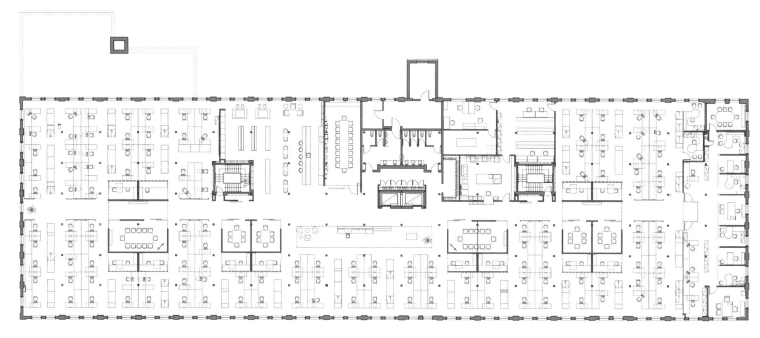

plan
平面图